SYSTEMS: APPROACHES, THEORIES, APPLICATIONS

EPISTEME

A SERIES IN THE FOUNDATIONAL,
METHODOLOGICAL, PHILOSOPHICAL, PSYCHOLOGICAL,
SOCIOLOGICAL AND POLITICAL ASPECTS
OF THE SCIENCES, PURE AND APPLIED

Editor: MARIO BUNGE
Foundations and Philosophy of Science Unit, McGill University

Advisory Editorial Board:

VOLUME 3

SYSTEMS:
APPROACHES, THEORIES,
APPLICATIONS

INCLUDING THE PROCEEDINGS
OF THE EIGHTH GEORGE HUDSON SYMPOSIUM
HELD AT PLATTSBURGH, NEW YORK,
APRIL 11–12, 1975

Edited by

WILLIAM E. HARTNETT

Dept. of Mathematics, State University College, Plattsburgh, N.Y., U.S.A.

D. REIDEL PUBLISHING COMPANY
DORDRECHT-HOLLAND / BOSTON-U.S.A.

Library of Congress Cataloging in Publication Data

George H. Hudson Symposium, 8th, Plattsburgh, N.Y., 1975.
    Systems: approaches, theories, applications.

    (Episteme; v. 3)
    Includes bibliographies and index.
    1. System theory – Congresses.   I.   Hartnett, William, E.   II.
Title.
Q295.G46 1975   003   77–22238
ISBN-13: 978-94-010-1241-6   e-ISBN-13: 978-94-010-1239-3
DOI: 10.1007/978-94-010-1239-3

Published by D. Reidel Publishing Company,
P.O. Box 17, Dordrecht, Holland

Sold and distributed in the U.S.A., Canada, and Mexico
by D. Reidel Publishing Company, Inc.
Lincoln Building, 160 Old Derby Street, Hingham,
Mass. 02043, U.S.A.

TABLE OF CONTENTS

PREFACE   IX

ACKNOWLEDGEMENTS   XI

NOTES ON THE CONTRIBUTORS   XIII

1. A CATEGORY-THEORETIC APPROACH TO SYSTEMS IN A
   FUZZY WORLD
   by Michael A. Arbib and Ernest G. Manes   1

   1. Machines in a Category   2
   2. Fuzzy Machines   16

2. PARALLELISM, SLIDES, SCHEMAS, AND FRAMES
   by Michael A. Arbib   27

   1. Parallelism   27
   2. Slides and Schemas   30
   3. Frames and Schemas   33
   4. Development   37
   5. More on Parallelism   39

3. THE FUNDAMENTAL DUALITY OF SYSTEM THEORY
   by E. S. Bainbridge   45

   1. Introduction   45
   2. Networks   47
   3. Duality   51
   4. Conclusion   60

4. TOWARDS A SYSTEMS METHODOLOGY OF SOCIAL CONTROL
   PROCESSES
   by Walter Buckley   63

5. STATES AND EVENTS
by Mario Bunge   71

    1. Introduction   71
    2. Properties and Predicates   73
    3. Definition of a State Function   76
    4. Law Statements   78
    5. Lagrangian Law Schemata   80
    6. State Spaces   82
    7. Law Statements and Transformation Formulas   86
    8. Events and Processes   89
    9. Event Space   91
   10. The Category of Events   92
   11. Concluding Remarks   94

6. UNDERSTANDING SOCIAL AND ECONOMIC CHANGE IN THE
UNITED STATES
by Jay W. Forrester   97

    1. System Dynamics   98
    2. Dynamics to Be Represented   104
    3. Social and Economic Issues   109
    4. Structure of the Model   114
    5. Status, Schedule, Procedure   118

7. PATTERN DISCOVERY IN ACTIVITY ARRAYS
by George J. Klir   121

    1. Introduction   121
    2. Sampling Procedure   129
    3. Evaluation of Masks   141
    4. Reduction of ST-Structures   153
    5. Conclusions   154

8. A PURPOSIVE BEHAVIOR MODEL
by Peter Milner   159

9. COMPLEXITY AND SYSTEM DESCRIPTIONS
by Robert Rosen   169

10. CONCERNS, COMMENTS, AND SUGGESTIONS
    by William E. Hartnett   177

    1. Educational Concerns   177
    2. Useful Mathematical Models   180
    3. Problems of Applied Mathematics   180
    4. Modeling   182
    5. State Modeling of Objects   189
    6. Questions   195

INDEX   199

PREFACE

For many years I have believed in a particular style of education for myself. The idea is to focus on matters that you want to learn about, find a modest amount of money, and then organize a symposium of those matters, inviting knowledgeable individuals to participate – and, by extension – to come and help with my education. The Eighth George Hudson Symposium held at Plattsburgh, New York on April 11–12, 1975 was another attempt on my part to learn something.

The ostensible reason for the Symposium was explained in the Announcement of the Symposium as follows:

Systems Theory is currently one of the exciting areas of intellectual activity, attracting persons from diverse disciplines. In fact, it has almost become the prototype of inter-disciplinary effort. As such, it needs the interchange of ideas, viewpoints, and opinions as a necessary condition for growth. This Symposium was convened to bring together a number of persons – some of them experts and some beginners – for two days of concentrated interaction on Systems Theory. The breadth of the interests of the invited speakers can be noted from their "home" disciplines but space limitations forestall any attempt to document their actual current interests which range from brain function to political institutions to technoethics. The speakers were chosen for their expository and interactive ability as well as for their work in Systems Theory and ample time has been allowed for discussion with them.

The Symposium provided some measure of local interchange of ideas, viewpoints, and opinions; this publication is the global counterpart. For the most part, the papers are those which were presented at the Symposium, in some cases revised by the authors in the light of reactions and subsequent discussions. The paper by Bunge was prepared for the Symposium but he was not able to attend. The papers by Forrester and by Bainbridge were added after the Symposium. E. Manes gave a talk entitled 'A categorical catalyst: new mathematical ideas in system theory' at the Symposium; by mutual agreement, no manuscript was prepared. The Arbib-Manes paper was background for the Manes paper.

No central theme was set for the Symposium and hence the editor is spared the task of showing (and the reader of accepting) how each paper fits in with the central theme. Each paper speaks for itself and despite the diversity of themes there are remarkable similarities of notions. One comment should perhaps be made about the Arbib-Manes paper — more precisely, about the title. In their paper — and elsewhere — fuzzy sets, fuzzy automata, fuzzy machines, and fuzzy categories abound. In each case, fuzzy is an adjective which adds something to the noun it modified. Reading again all of the papers of this volume, I am convinced that "fuzzy" as in "fuzzy world" is redundant. One might almost be tempted to say that the world is inherently fuzzy. If such is indeed the case, I hope that this volume will help us deal with the fuzz.

Plattsburgh, New York, U.S.A.   WILLIAM E. HARTNETT
March 1976

ACKNOWLEDGEMENTS

To D. Reidel Publishing Company and to Jaakko Hintikka, Editor of *Synthese*, for permission to reprint 'A Category-Theoretic Approach to Systems in a Fuzzy World', Michael A. Arbib and Ernest G. Manes, *Synthese* **30** (1975) 381–406.

To Jay W. Forrester for permission to print copyrighted material.

To the Faculty of Science and Mathematics, State University College, Plattsburgh, New York, for funding the Eighth George Hudson Symposium on Systems: Approaches, Theories, Applications, held at Plattsburgh on April 11–12, 1975.

To Mario Bunge and to the Canada Council. To Bunge for having me as a Research Associate during a very educational year and to the Council for underwriting my weekly visits to McGill University. Their support provided the gestation period for my paper.

To Carol Burnam, Secretary of the Department of Mathematics, for many things, particularly for the loving care and attention she has given to my work over the years. I am especially grateful for all the onerous tasks she has undertaken in connection with the Eighth George Hudson Symposium and the publication of this volume.

W. E. H.

NOTES ON THE CONTRIBUTORS

Michael Arbib is Professor of Computer Science and of Psychology at University of Massachusetts, Amherst, Massachusetts, U.S.A.

E. Stuart Bainbridge is Assistant Professor of Mathematics at University of Ottawa, Ottawa, Ontario, Canada, and is currently Visiting Professor at Columbia University.

Walter Buckley is Professor of Sociology at University of New Hampshire, Durham, New Hampshire, U.S.A.

Mario Bunge is Head of the Foundations and Philosophy of Science Unit at McGill University, Montreal, Quebec, Canada, and is currently Visiting Professor at Instituto de Investigaciones Filosóficas, Universidad Nacional Autónoma de Mexico.

Jay W. Forrester is Germeshausen Professor at Massachusetts Institute of Technology, Cambridge, Massachusetts, U.S.A.

William E. Hartnett is Professor of Mathematics at State University of New York at Plattsburgh, New York, U.S.A.

George Klir is Professor at the School of Advanced Technology of the State University of New York at Binghamton, New York, U.S.A., and is currently Visiting Fellow at the Netherlands Institute for Advanced Studies in Wassenaar, The Netherlands.

Ernest Manes is at the Department of Mathematics, University of Massachusetts, Amherst, Massachusetts, U.S.A.

Peter Milner is Professor of Psychology at McGill University, Montreal, Quebec, Canada.

Robert Rosen is Killam Professor of Biomathematics in the Department of Biophysics and Biomathematics at Dalhousie University, Halifax, Nova Scotia, Canada.

CHAPTER 1

A CATEGORY-THEORETIC APPROACH TO SYSTEMS IN A FUZZY WORLD*

MICHAEL A. ARBIB AND ERNEST G. MANES

The last 30 years have seen the growth of a new branch of mathematics called CATEGORY THEORY which provides a general perspective on many different branches of mathematics. Many workers (see Lawvere, 1972) have argued that it is category theory, rather than SET THEORY, that provides the proper setting for the study of the FOUNDATIONS OF MATHEMATICS.

The aim of this paper is to show that problems in APPLIED MATH-EMATICS, too, may find their proper foundation in the language of category theory. We do this by introducing a number of concepts of SYSTEM THEORY which we unify in our theory of MACHINES IN A CATEGORY. We write as system theorists, not as philosophers. Our hope is to stimulate a dialogue with philosophers of science as to the proper role for category theory in a systematic analysis of a fuzzy world. We do not discuss applications to biology or psychology – the framework presented here is at a very high level of generality, and does not address the particularities which give these disciplines their distinctive flavor.

This paper is divided into two Sections. In Section I, we sketch how the subjects of control theory, computers and formal language have grown out of the urdisciplines of MECHANICS and LOGIC; and then present the formal concepts of *sequential machine*, *linear machine*, and *tree automaton*. We show how our notion of MACHINE IN A CATEGORY provides an uncluttered generalization of these three concepts.

In Section II, we introduce the 'fuzzy world'. Although the study of quantum mechanics provides the best known framework, we stay within system theory, showing how PROBABILITY, MECHANICS and LOGIC gave rise to the study of markov chains, structural stability and multi-

* The research reported in this paper was supported in part by the National Science Foundation under Grant No. GJ 35759.

W. E. Hartnett (ed.), Systems: Approaches, Theories, Applications, 1–26.

valued logics. We then present the formal concepts of *nondeterministic sequential machine, stochastic automaton* and *fuzzy-set automaton*. Our notion of FUZZY MACHINE will generalize all three. Of particular interest will be the demonstration that, although fuzzy machines *generalize* machines in a category, we can – by a suitable enlargement of viewpoint – regard them as a special case.

The paper is self-contained both as to system theory and to category theory – but many topics must be but briefly outlined in an expository paper of this kind. The reader wishing a fuller *introduction* to category theory is referred to our book (Arbib and Manes, 1975); a text on control theory is Athans and Falb (1966); for system theory see Padulo and Arbib (1974) and Kalman *et al.* (1969); many other concepts of machine theory appear in Bobrow and Arbib (1974); our theory of machines in a category appeared in Padulo and Arbib (1974), Bobrow and Arbib (1974), Arbib and Manes (1974a), while the technical details of fuzzy machines appear in Arbib and Manes (to appear). The state of the art in applying category theory to systems and automata is reflected in Manes (1975b).

1. MACHINES IN A CATEGORY

In Figure 1, we schematize the evolution of Machines in a Category from concepts in generalized mechanics and formal logic through the study of control theory, the impact of computers, and notions of formal linguistics. The paragraphs below are lettered with the arrows they describe:

A   Building on the work of Newton and its refinement by such workers as Legendre, Hamilton, in the middle of the 19th Century, gave the following formulation of generalized mechanics: The vector of *generalized positions*, $q = (q_1, ..., q_n)$, one for each degree of freedom of the system must be augmented by $p = (p_1, ..., p_n)$, the vector of *generalized momenta*, one for each degree of freedom of the system. There is then a function $H(p, q)$, the *Hamiltonian*, of these variables, in terms of which we may express the system dynamics:

$$\begin{cases} \dot{q}_j = \dfrac{\partial H}{\partial p_j} & \text{for } 1 \leqslant j \leqslant n \\[2ex] \dot{p}_j = -\dfrac{\partial H}{\partial q_j} & \text{for } 1 \leqslant j \leqslant n. \end{cases}$$

Thus, with Hamilton we see very vividly that we may study systems which are described by *the evolution of state vectors over time*, with this evolution governed by vector differential equations of the form

$$\dot{q} = f(q)$$

where now the state q includes position, momentum, and any other relevant variables as components.

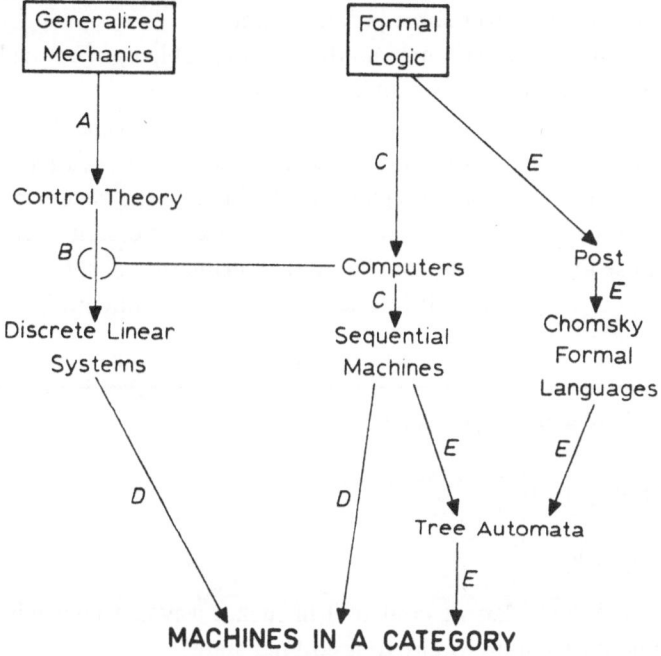

Fig. 1.

The transition to control theory comes when we emphasize that the differential equation describing the evolution of the state of a system contains a number of parameters, representing forces, which can be manipulated from outside the system, so that we may write down the change of state as a function

$$\dot{q} = f(q, x) \tag{1}$$

not only of the *state vector* itself, as in the classical formulation, but also as a function of a *control vector* x. We should also note that only certain aspects of the state will actually be measurable at any time, so that we may introduce an output vector *y* which is a function

$$y = \beta(q) \tag{2}$$

of the instantaneous state. For example, in classical mechanical systems, we can observe only the positions 'instantaneously', while the momenta – or the related velocity variables – must be built up from observation of changes in position over some period of time.

We now turn (Box 1) to three mathematical problems of control theory, which underpin the central problem of *optimization*.

Given a system described by a pair of equations giving (1) the rate of change of the state and (2) the observable output as a function of the state, we are to find a control signal which will drive the system from some initial state to a desired final state in the quickest possible way, or with the least use of energy – as, for example, of firing the rockets of a satellite in such a way as to bring it into a desired stable configuration. Clearly,

---

Three Problems of Control Theory

Given a system $\begin{cases} \dot{q} = f(q, x) \\ y = \beta(q) \end{cases}$

we may ask:

Is it **reachable**? Can we control it in such a way as to drive it from some initial state to any desired final state?

Is it **observable**? Given the system in an unknown state, can we conduct experiments upon it (apply controls, measure outputs) in such a way as to eventually determine the system's state.

Given a system whose equations are unknown, the **realization problem** is to determine a set of states, a dynamics *f*, and an output function *β* which correctly describe the observed input-output behavior of the system.

Box 1

however, before we analyze what is the most efficient way to bring it into position, we must know whether any suitable control exists at all, and this is the question of *reachability*. [Incidentally, it is worth noting that optimal control is closely based on the work of Hamilton, for Hamilton had observed that the trajectory of a system following given laws of motion was such as to minimize the value of a certain function. It is a natural transition, then, to apply these techniques to seek an input – or control – trajectory which will minimize some evaluation of the cost or time of system performance, and this approach is the basis of Pontryagin's maximum principle, one of the fundamental techniques of optimal control.]

If reachability is an important question in the design of feedback control systems – given a state, does there exist a control we can apply to move the system from that state to some other, desired, state – then no less important a question must be the one of *observability*. We have already commented that the instantaneous output of the system will in general tell us only some small portion of what we need to know about its state. But feedback control usually requires that we know all of the state before we can determine what is the proper input to apply. Thus, it is our concern to determine when a system is observable: namely, we wish to know how, given the system in an unknown state, we can conduct experiments upon it – namely by applying controls and measuring the consequent outputs – in such a way as to eventually determine the system's current state. Thereafter, our knowledge of the dynamics will allow us to update the state as we apply the appropriate controls to its behavior.

The above prescription is based upon our knowing the Equations (1, 2) which govern the system. This of course raises the very realistic problem of how we might find these equations in the first place. In general, if we come upon a system to which we can apply certain inputs, and for which we can observe certain outputs, we wish to determine a state-space which can mediate the relationship between the inputs and the outputs, and we then wish to determine the dynamics and the output function which correctly describe the observed input/output behavior of the given system. This is the *realization problem*, and we are frequently concerned to find a realization which is minimal in the sense of having the smallest state-space possible. One of the most pleasing general results of control theory is that if a realization is indeed minimal, then it must be both reachable and observable.

B   However, the treatment of arbitrary systems described by differential equations is too complex for efficient mathematical solution. One of the most common ways of approximating a complex system is by using linear equations. Moreover, the advent of the computer as the tool par excellence for controlling a system has led us to move from *continuous time* systems described by differential equations to *discrete time* systems in which we *sample* the behavior of the system, and apply inputs, at regular intervals, so that we describe the system in terms of equations which show how it changes from one sampling period to the next. In fact by using an approximation to the rate of change predicted by the derivative, and by using Taylor series, we can come up with a linear approximation to the change in state of the system over the sampling period which is linear, and we may also approximate the output by a linear function of the state:

$$q(t + \varDelta t) \doteqdot q(t) + f(q(t), x(t)) \, \varDelta t$$

$$\doteqdot q(t) + \frac{\partial f}{\partial q} \cdot q(t) \, \varDelta t + \frac{\partial f}{\partial q} \cdot x(t) \, \varDelta t$$

(3) $$= F q(t) + G x(t)$$

where

$$F = \left[I + \varDelta t \, \frac{\partial f}{\partial q} \right]; \qquad G = \varDelta t \, \frac{\partial f}{\partial x}.$$

$$y(t) \doteqdot \frac{\partial \beta}{\partial q} \cdot q(t)$$

(4) $$= H q(t)$$

where

$$H = \frac{\partial \beta}{\partial q}.$$

It is an *empirical fact* that many control systems can be *usefully approximated* by descriptions of the form (3)/(4) using *constant matrices F, G* and *H*.

C   If computers encouraged the passage from general differential equations to discrete linear systems – or *linear machines* as we will call them from now on – they also gave rise to new discrete systems in their own right, which in no way were to be considered as approximations to continuous systems. The concepts of truth values in a two-valued logic which could be computed upon in a numerical-like but non-numerical way, due to George Boole, provided the proper framework in the 1930's and 1940's for the development of a formal theory both of relay switching networks and the McCulloch-Pitts theory of formal networks. These led to the general theory of *sequential machines*, which – among other things – provided the proper formal framework for talking about the various subsystems of a computer. For example (Figure 2) we can describe a vending machine which accepts nickels, dimes and rests – the set of inputs is $X_0 = \{n, d, R\}$. It vends a candy bar, C, when 15¢ has been received from the initial state, puts out 20¢ if it has received either 2 dimes or 2 nickels and a dime starting from the initial state, and otherwise emits nothing, $\emptyset$ – so that the output set is $Y = \{\emptyset, C, 20¢\}$. The current state and current input determine the next state via a function δ – an arrow leads from node q via arrow x to node $\delta(q, x)$. The current output is a function β of the current state – we mark the node for state q with the notation $q/\beta(q)$.

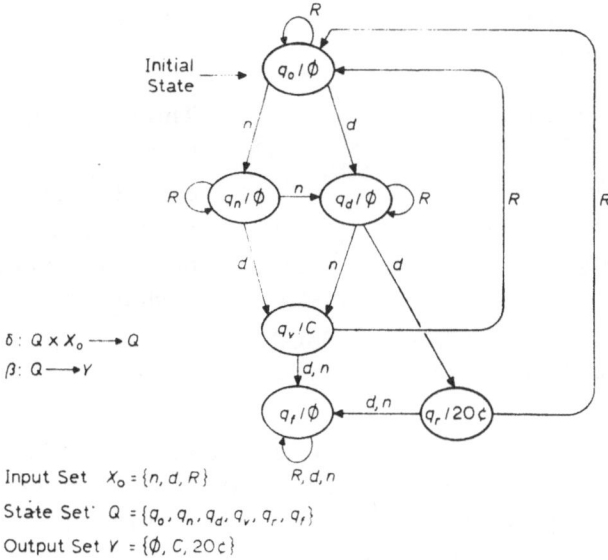

Fig. 2. The 15¢ machine.

D   The point to stress here is that the various input, state, and output sets involved here are small finite sets, and are in no way the Euclidean spaces of linear system theory. In fact (Box 2) we may see that the theory of sequential machines and the theory of linear machines live in quite different domains of discourse:

First, let us examine sequential machines. It is common to assign to each machine an initial state – in this case we have represented that initial state by the map τ from the one-element set 1 to the state set Q whose image is precisely the initial state q_0. The dynamics $\delta: Q \times X_0 \to Q$ is then a map which assigns to each state and each input of the sequential machine the state into which it will next settle, whereas the output map $\beta: Q \to Y$

Formal Definitions

Sequential Machines	**Linear Machines**
Initial State	Input Map
$\tau: 1 \to Q$	$G: I \to Q$
Dynamics	Zero-Input Dynamics
$\delta: Q \times X_0 \to Q$	$F: Q \to Q$
Output Map	Output Map
$\beta: Q \to Y$	$H: Q \to Y$
This lives in the category **Set**:	This lives in the category **Vect**:
each *object* is a set; each *morphism* (arrow) is a map.	each *object* is a vector space; each *morphism* (arrow) is a *linear map*.

Box 2

---

assigns to the current state the corresponding output. We stress that sequential machines live in the category **Set** – a domain of mathematical discourse comprising sets and arbitrary maps between those sets.

In describing a linear machine, we give an input map $G: I \to Q$, a zero input dynamics which is simply the map F from the state set Q into itself, and an output map $H: Q \rightarrow Y$. These describe the behavior of the machine via $q(t + \Delta t) = Fq(t) + Gx(t)$; $y(t) = Hq(t)$. The appropriate domain of discourse here is the category **Vect** in which now the objects are vector spaces and each morphism – i.e., arrow going from one object to another – is a linear map .[We have lined up elements of the definitions of sequential machines and linear machines in Box 2 in a way that will seem mysterious to the reader. We hope that the reason will become clear by inference from our general definition of machines in a category in Box 4 below.]

Clearly, at this stage it is proper that we admit that the notion of a CATEGORY or mathematical domain of discourse implicit in our above comparison is in fact a formal concept of mathematics. In fact, we have as the basic notions of category theory the idea of a category and of a functor (Box 3).

A *category* $\mathscr{K}$ is a domain of mathematical discourse in which we have a collection of *objects*, such as the arbitrary sets of **Set** or the vector spaces of **Vect**, together with, for each pair A, B of objects, a collection $\mathscr{K}(A, B)$ of *morphisms* from the first to the second – these correspond to the arbitrary maps of one set into another of **Set**, or the linear maps from one vector space into another of **Vect**. As in both of these examples, we may compose morphisms so long as the first ends where the second begins – and the composition is *associative*, i.e., we may string together an arbitrary number of composable maps and know that the overall composition is uniquely defined, irrespective of the 'bracketing' of the constituent morphisms. Moreover, we may associate with each object an identity morphism – this corresponds to the map which sends each element to itself in **Set** and **Vect** – which has the property that if we compose it with any other morphism, the result is that other morphism. Incidentally, this equivalent definition of the identity map exemplifies the difference between the set theory (define everything in terms of elements) and the category theory (define everything in terms of morphisms) approach to the foundations of mathematics.

So far, so good. A somewhat more technical concept basic to any use of the language of category theory is that of a *functor*. Briefly put, a functor is simply a passage from one category to another in such a way that the identities, and the composition of morphisms, are respected. In particular, a very useful idea in category theory has been that of 'chasing commutative diagrams' – drawing graphs in which morphisms take us from one

Basic Notions of Category Theory

A Category $\mathscr{K}$ **is a domain of mathematical discourse comprising**
a *collection of objects*
for each pair A, B of objects a collection $\mathscr{K}(A, B)$ of *morphisms*

$$f: A \to B \text{ or } A \xrightarrow{f} B$$

with *domain A* and *codomain B*
together with a law of composition

$$g \cdot f: A \to C = A \xrightarrow{f} B \xrightarrow{g} C$$

which is associative and has identities $id_A: A \to A$.

A functor H from category $\mathscr{K}$ to category $\mathscr{L}$

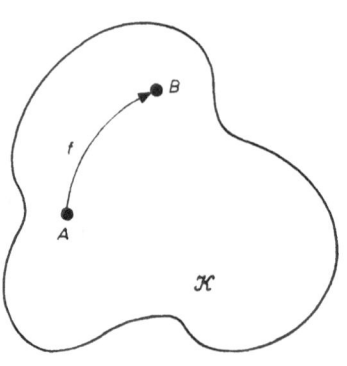 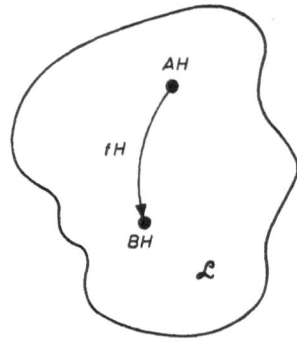

sends	*to*
objects A	objects AH
morphisms $f: A \to B$	morphisms $fH: AH \to BH$
in $\mathscr{K}$	in $\mathscr{L}$

in a 'nice' way, namely

If $f = id_A: A \to A$ then $fH = id_{AH}: AH \to AH$
If $f = A \xrightarrow{g} B \xrightarrow{h} C$ then $fH = AH \xrightarrow{gH} BH \xrightarrow{hH} CH$.

Box 3

object to another over diverse paths in such a way that the overall composition is the same. E.g., to say that

(5)

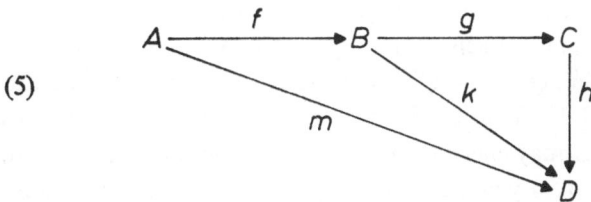

commutes, is to say that $k \cdot f = m$, $h \cdot g = k$ and $h \cdot g \cdot f = k \cdot f = m$. The iterated application of the fact that a functor preserves identities and composition allows us to easily deduce that it must also preserve the commutativity of any diagram – i.e., that if we replace each object A by the object AH and if we replace each morphism f by the morphism fH, then if different paths from one initiation point to one termination have the same composition in the original diagram, then they must have equal compositions in the transformed diagram. For example, if (5) commutes in $\mathscr{K}$ then

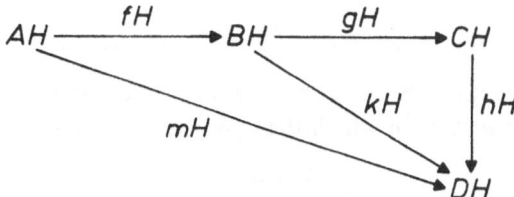

commutes in $\mathscr{L}$ – e.g., $mH = (k \cdot f) H = kH \cdot fH$.

    With these concepts before us we can now present *the key concept of machines in a category* $\mathscr{K}$. We should *not*, as we were encouraged to do in the theory of sequential machines, think of the input of a machine as being a set – or, more generally, an object – of inputs. Rather, we should *think of the* **input** *as being a* **process** which transforms the state object Q into a new state object QX. In all cases, we are to think of X as being a functor from the given category $\mathscr{K}$ to itself. Then, given this object QX upon which the dynamics is to act, a *dynamics* is simply a $\mathscr{K}$-morphism $\delta: QX \rightarrow Q$.

    Returning to Box 2, we see that for sequential machines, the category

$\mathscr{K}$ is **Set**, and the functor X transforms a state set Q into the cartesian product $Q \times X_0$ of all state-input pairs; while in the case of linear machines we work in the category $\mathscr{K} = $ **Vect**, and our functor X leaves things unchanged so that $QX = Q$. [To see that these really are functors, we must show how they act on morphisms. For $f: Q \to Q'$ in **Set** and $X = - \times X_0$: **Set** $\to$ **Set**, we define $fX: Q \times X_0 \to Q' \times X_0$ to send (q, x) to $(f(q), x)$. For $f: Q \to Q'$ in **Vect** and $X =$ identity: **Vect** $\to$ **Vect**, we define $fX: Q \to Q'$ to be simply f. The reader may check the functor conditions of Box 3.] Then, a sequential machine has dynamics $\delta: Q \times X_0 \to Q$, while linear machines have dynamics $F: Q \to Q$. With this we see that both types of machine of Box 2 are subsumed in our general notion of MACHINE IN A CATE-GORY, summarized in Box 4. Summarizing, we see that a machine in a category requires us to specify a functor X from $\mathscr{K}$ into itself which is a

---

MACHINES IN A CATEGORY

X-Machines $\begin{cases} \tau: & I \to Q \\ \delta: & QX \to Q \\ \beta: & Q \to Y \end{cases}$

$X: \mathscr{K} \to \mathscr{K}$ is a *functor*; τ, δ and β are $\mathscr{K}$-morphisms

We stress that input is a **process** *which converts the state-object Q into a new object QX* on which the dynamics can operate

Box 4

---

process which converts the state object Q into a new object QX on which the dynamics δ can operate. We must specify a $\mathscr{K}$-morphism τ from I to Q – in the case of sequential machines this gives us the initial state, while it gives the input map of a linear machine. Finally, we give a morphism β from Q to Y – which provides an output map in both cases.

E   Instead of giving a formal treatment, let us just briefly note that tree automata do indeed fit into this general framework of machines in a category. Here, we briefly note that Post's theory of canonical systems was specialized by Chomsky to yield his formal theory of languages, and that many authors soon realized that the appropriate theory for handling the derivation trees of formal linguistics was the theory of *tree automata*, which could be seen as a straightforward generalization of the theory of sequential machines we have discussed above. Rather than give the general definition of tree automata, however, let us content ourselves with a simple example (Figure 3) of processing binary arithmetic trees. Here we start at the bottom – at the 'leaves' – and combine pairs of numbers by addition and multiplication until finally at the 'root' of the tree we have the overall evaluation of the arithmetic expression represented by the tree. Let us see how we can think of this as a machine in a category in the sense of Box 4. Here we are to think of the state set as being the set **N** of all natural numbers, and we now introduce a functor X: **Set** → **Set** on the

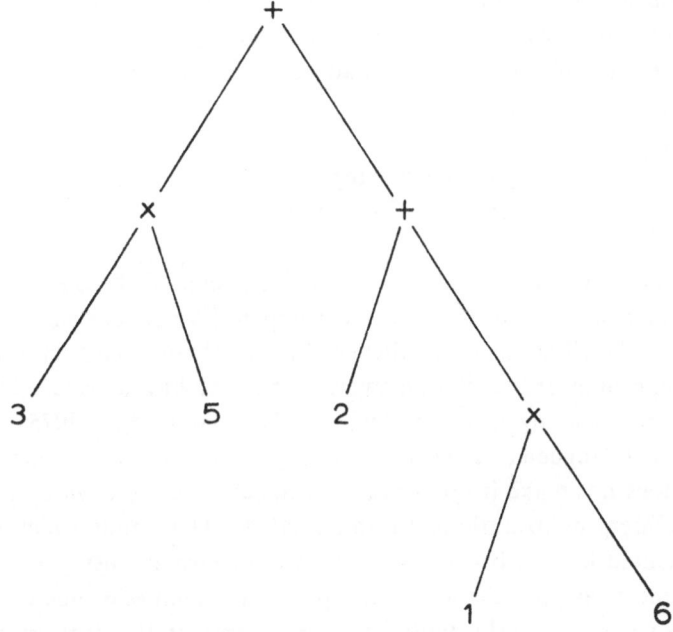

Fig. 3. Processing binary arithmetic trees. State Set Q is N (the natural numbers) in this example. Introduce a functor X: **Set**→**Set**

$$QX = Q \times Q \times \{+\} \cup Q \times Q \times \{\times\}.$$

Then a map δ: $QX \rightarrow Q$ gives the dynamics:

$$\delta(q_1, q_2, +) = q_1 + q_2$$
$$\delta(q_1, q_2, \times) = q_1 \times q_2.$$

category of sets, which sends each state set Q to the union QX of two sets, one being $Q \times Q \times \{+\}$ and the other being $Q \times Q \times \{\times\}$. We then see that a map from QX to Q gives us precisely the two maps we need to evaluate nodes of the tree as we pass from the leaves to the root.

With this successful subsumption of tree automata in a framework designed to embrace sequential machines and linear machines, we have almost completed the first part of the paper. But, before we look at what happens to this theory in a 'fuzzy world', it seems worthwhile to quickly summarize a number of results which have been obtained in the theory of machines in a category, even if we do not have space to spell out any of the details. In fact, given any functor X from the category $\mathscr{K}$ into itself we can define a category $\mathbf{Dyn}(X)$ of X-*dynamics* – the objects are precisely the X-dynamics, while a $\mathbf{Dyn}(X)$-morphism – or a *dynamorphism* – is a $\mathscr{K}$-morphism of state objects which 'respects' the dynamics – we might either apply the dynamics and code the resulting state, or we may code QX and then apply the second dynamics – the result is the same, as expressed in the commutative diagram

$$\begin{array}{ccc} QX & \xrightarrow{\delta} & Q \\ {\scriptstyle hX}\downarrow & & \downarrow{\scriptstyle h} \\ Q'X & \xrightarrow{\delta'} & Q' \end{array}$$ $\mathbf{Dyn}(X)$ is a **category** because X is a **functor**

This category is the setting for the major results of the theory of machines in a category which we have developed. [We should also mention that other contributions to the theory of machines in a category – though not using exactly the same framework as that we have developed here – have been made by Goguen (1972, 1973), Bainbridge (1975), Ehrig *et al.* (1974), Goguen *et al.* (1973) and others. However, the nature of our survey does not make it appropriate to indicate here the ways in which these different contributions are interrelated.] The results which follow are presented far too briefly to allow comprehension – using as they do the technical category-theoretic concept of an adjoint of a functor. However, the very point of this tantalizingly brief presentation is to stress how important adjoints are to system theory; and we hope that many readers will be tempted to turn to Arbib and Manes (1975), Padulo and Arbib (1974), Bobrow and Arbib (1974), and Arbib and Manes (1974a).

We introduce a new functor $U\colon \mathbf{Dyn}(X) \to \mathscr{K}$ which sends an object (Q, δ) of $\mathbf{Dyn}(X)$ to Q in $\mathscr{K}$, and sends a dynamorphism $h\colon (Q, \delta) \to$

$\rightarrow (Q', \delta')$ to the underlying $\mathscr{K}$-morphism $h: Q \rightarrow Q'$. We call it the *forgetful functor* because it 'forgets' the dynamics δ and just remembers the underlying state-object Q.

Category theorists give a central role to the notion of *adjoint of a functor*. In some circumstances we may associate to a functor $H: \mathscr{L} \rightarrow \mathscr{K}$ another functor $F: \mathscr{K} \rightarrow \mathscr{L}$ called the *left adjoint* of H. In other circumstances, there exists a functor $G: \mathscr{K} \rightarrow \mathscr{L}$ called the *right adjoint* of H. The definition of adjoints is beyond the scope of this paper (see Arbib and Manes, 1975, Chapter 7 for the details), but we note the terminology that if H has left adjoint F and B is an object in $\mathscr{K}$, then we say that BF is the **free** $\mathscr{L}$-object *over* B; while if H has right adjoint G, we say that BG is the **cofree** $\mathscr{L}$-object *over* B. With this terminology we may summarize some of our results:

First, we showed that if the forgetful functor $U: \mathbf{Dyn}(X) \rightarrow \mathscr{K}$ from the category of X-dynamics to the underlying category $\mathscr{K}$ has a left adjoint $F: \mathscr{K} \rightarrow \mathbf{Dyn}(X)$ – so that we may talk of *free dynamics QF* in $\mathbf{Dyn}(X)$ – then we can in fact construct a reachability theory and a theory of minimal realization. This theory includes sequential machines, linear machines, tree automata, and many other examples.

If on the other hand we require that the forgetful functor has a right adjoint $G: \mathscr{K} \rightarrow \mathbf{Dyn}(X)$ – so that we may construct a *cofree dynamics QG* in $\mathbf{Dyn}\,X$ – we are then able to construct an observability theory and a cominimal realization theory – which is much the same as a minimal realization theory, with differences that are too technical to detain us here. In any case, we find that tree automata do not correspond to functors X which yield forgetful functors with right adjoints, but sequential and linear machines do. Thus, both sequential and linear machines are examples of machines in a category for which the corresponding forgetful functor has both a left and a right adjoint, and we have found that in this case we get an exceptionally simple minimal realization theory using what are called image factorizations, and that we also have a framework for studying duality of systems based upon the fundamental concept of categorical duality (Arbib and Manes, 1975). In particular, of course, we may talk about both reachability and observability for such systems. To further tantalize the reader, we point out that, for I as in Box 4, IF is the 'object of input experiments'. Since IF is determined uniquely by X (Arbib and Manes, 1975, p. 113), the nature of 'input experiments' is not determined independently by intuition – a new principle in system theory. This principle has surprising consequences for affine machines (Goguen, 1972) and group machines (Arbib and Manes, 1974b).

Summarizing, then, we have seen that with the idea of a functor we can embrace a far larger class of automata than we can by restricting ourselves to the situation in which the dynamics must act on something with the form of $Q \times X_0$; and – as the above flash-through of results indicates – the category theory concept of adjoints of functors is central to our approach to general system theory.

We reiterate that the above survey is far too brief, but it should be sufficient to set the stage for the new perspective that is required when we start looking at different aspects of nondeterminism in our approach to systems in a fuzzy world.

2. FUZZY MACHINES

We have now seen how to use category theory to provide a general perspective (Figure 1 to Box 4) for a number of apparently disparate classes of systems: sequential machines, linear machines, and tree automata. But the time has come to face up to the fact that we live in a 'fuzzy' world – there is no guarantee that we can be sure of the next state of a system in the real world. In the rest of this paper, we are going explore a somewhat paradoxical approach to the 'fuzziness', namely that in which one can give a precise prescription of the range of possibilities for the next state from any given starting condition. (But we emphasize at once that we will axiomatize a class of such prescriptions, frankly recognizing that there are many different kinds of fuzziness.)

The first way in which nondeterminism entered the world of automata theory was through the study of nondeterministic sequential machines (F of Figure 4). This was in part motivated by the study of formal languages – for in designing machines to parse a sentence one had to be aware of the fact that the initial portion of a sentence could be consistent with a number of possible parsings, so that there was no unique way to classify the next word, but rather a number of possible ways consistent with the information already processed. In any case, whatever the history, there has become entrenched the idea of a *nondeterministic sequential machine* – we suggest that perhaps a better word would be 'possibilistic' –

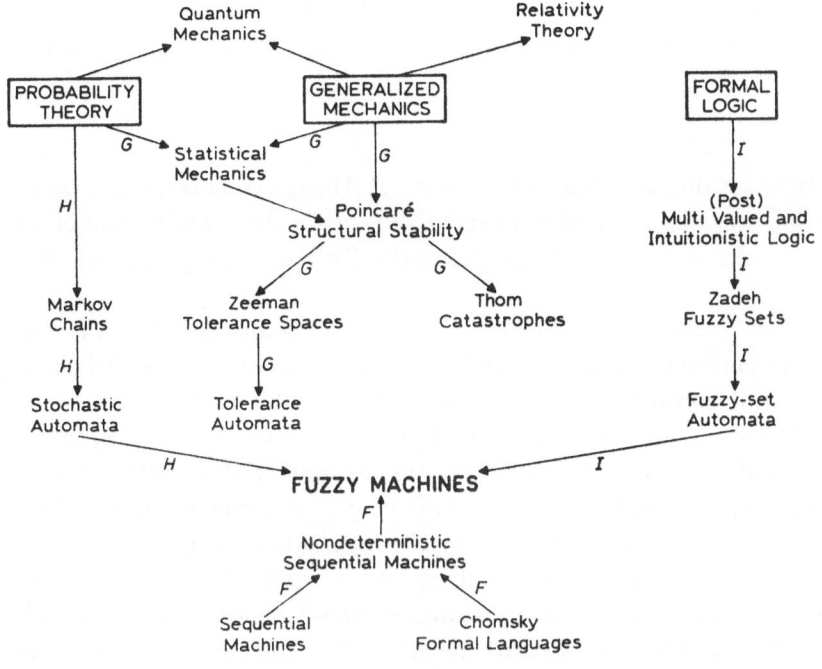

Fig. 4.

in which the current state and the input do not determine a single unique next state, but rather determine a set of *possible* next states, so that the dynamics maps the set of (state, input) pairs into an element of 2^Q, the set of subsets of the state set Q.

$$\delta: Q \times X_0 \to 2^Q.$$

The idea, then, is that in any run of the machine, one and only one state will appear at any given time, but if state q appeared at time t and input x were then applied, the state at time $t+1$ *must* belong to the set $\delta(q, x)$ of states.

Now, we may observe that the passage from Q to 2^Q is the object map of a functor of the category **Set** into itself,

$$2^{(-)}: \textbf{Set} \to \textbf{Set} \text{ is a functor}$$
$$Q \mapsto 2^Q$$
$$[f: Q \to Q'] \mapsto [2^f: 2^Q \to 2^{Q'}: S \subset Q \mapsto f(S) =$$
$$= \{f(s) \mid s \in S\} \subset Q'].$$

This suggests that the nondeterministic sequential machines we have just looked at may be considered to be a special case of dynamics expressed in the form

$$\delta\colon QX \to QT$$

for some suitable choice of a functor T. The question before us, then, is what are suitable restrictions on functors T for the consideration of such dynamics to be in fact the proper setting for 'dynamics in a fuzzy world'?

G   Before we turn to this rather technical question, however, it is worth continuing the historical perspective of Figure 1 by considering, in Figure 4, various ways in which the idea of a 'fuzzy world' has been approached. Of course, this historical view of ours is a very sketchy one, and we can only hope that some more careful historian or philosopher of science will take this lead to more carefully chart the interconnections between these ideas. In any case, let us briefly notice that generalized mechanics in the classical sense has recently spawned two most important new theories of mechanics, namely quantum mechanics (with crucial use of probability theory) and relativity theory. Unfortunately, we have nothing further to say at this time about these important developments, but wish to draw attention briefly to the fact that classical mechanics and probability theory have also given rise to statistical mechanics – namely the description of large systems in terms of the average behavior of their myriad deterministic (or possibly quantum mechanical) components.

The theory of statistical mechanics is still in an unsatisfactory form, and we believe that its proper development is one of the great challenges of system theory. Here, however, let us briefly note that Poincaré, in pondering the various problems of celestial mechanics, came up with a very crucial notion of structural stability – a notion very much appropriate to the conduct of scientific study in a fuzzy world. Briefly, he noted that in taking any system, it is not possible to determine the parameters of that system with complete exactitude. It is thus, then, a matter of crucial import that no very delicate change in the parameters of the system should drastically alter its behavior – for then we could have confidence in the predictions that were made. This, then, is the idea of a *structurally stable system*: a system whose behavior is only changed slightly by a slight change in the parameters that describe the equations of motion of that system. Interestingly, these ideas of Poincaré have led to two recent developments. One is Thom's theory of *catastrophes* (1972) – in which Thom classifies those parameters of system description which lie at the borderline between two different domains of structural stability. It is perhaps worth noting in passage our belief that Thom's mathematical contributions here are of vital importance to system theory; while at the same time expressing a measure of skepticism about the way in which Thom has suggested that his theory of catastrophes has immediate applications to such diverse fields of applied mathematics as theoretical embryology and linguistics.

A more direct descendent of Poincaré's ideas is the theory of tolerance spaces due to Zeeman, in which he replaced the idea of a topology on a space by the more discrete notion of a tolerance: namely a reflexive and symmetric relation which tells us of any two points of the space whether or not they are in tolerance of one another. This then suggested to Arbib the idea of a tolerance automaton – namely a sequential machine in which the dynamics and output are 'continuous' with respect to tolerances on the various spaces involved. It has recently been noted by Dal Cin that we may make such tolerance automata into machines in a category in a fairly obvious way.

With this, then, let us turn to the remaining two evolutions in Figure 4 – namely, that from the Markov chains developed by the probability theorists; and that which we may recognize as part of the evolution of multivalued and intuitionistic logics (the name of Post occurs here as well as in the canonical systems which led to formal language theory) from classical Boolean logic.

H Markov chains were developed in the late 1800's as a way of modelling the dynamics of a classical system for which one could at best give probabilities as to the next state given the present state, rather than the classical systems with which we started our discussion in this paper in which the current state determined the future states for all time. The stochastic automaton, then, is related to Markov chains just as our control systems are related to classical mechanical systems. Namely, we introduce a set of inputs, such that for each input there is a corresponding Markov chain, with the probability distribution of the next state being determined by the Markov chain indexed by the current input.

More formally, a *Markov chain M* is given by a set $\{q_1, ..., q_n\}$ of states and an $n \times n$ stochastic matrix $P = (P_{ij})$ whose interpretation is that if M is in state q_j at time t, then it will be in state q_i at time $t+1$ with probability p_{ij}. A *stochastic automaton* has its dynamics given by a set $X = \{x_1, ..., x_m\}$ and a collection of m Markov chains, one P^x for each input $x \in X_0$. If M is in state q_j at time t and receives input x, then it will be in state q_i at time $t+1$ with probability p_{ij}^x. Here the dynamics is

$$\delta: Q \times X_0 \to Q\mathbf{P}:(q_j, x) \mapsto \begin{bmatrix} p_{1j}^x \\ \vdots \\ p_{nj}^x \end{bmatrix}$$

where $\mathbf{P}: \mathbf{Set} \to \mathbf{Set}$ is a functor with

$Q\mathbf{P} =$ set of probability distributions on Q (If $p \in Q\mathbf{P}$, let $p(q)$ denote the probability of q)

$$f\mathbf{P}: Q\mathbf{P} \to Q'\mathbf{P}: f\mathbf{P}(p): q' \mapsto \sum_{q \in f^{-1}(q')} p(q).$$

We see once again that the dynamics is of the form $QX \to QT$, where now T is the functor $P: \mathbf{Set} \to \mathbf{Set}$ which sends a set Q to the set of all probability distributions on Q.

I For our last example of a functor T for our general theory, we turn to *fuzzy sets*. This notion seems to have been independently established by Zadeh (1965), although it is clearly a special case of ideas developed by many authors in looking at multivalued and intuitionistic logic. Briefly, Zadeh observed that there are many 'sets' in the world for which one cannot make the confident assertions of membership or nonmembership demanded by classical set theory. For example, the set of all 'tall people' is such a set. Certainly someone who is three feet tall does not belong to the set, while someone who is seven feet tall certainly does. But what of someone 5'3" tall? Perhaps they almost belong to it, say with 'weight' 0.3, while someone of height 5'8" might belong to the set with membership strength 0.8. On this basis, then, Zadeh defines a fuzzy set in the universe W to simply be a map A from W to the continuous interval [0,1] of real numbers, with $A(w)$ being the strength of membership of w in A.

Before going further, it is perhaps worth noticing that there is a certain horror in this approach to the problem of fuzziness – for if it seemed unreasonable to simply say of any element whether or not it belonged to the set of tall people, surely it seems even more unreasonable in this fuzzy world of ours to assign so precise a number as 0.7 to membership. It may perhaps be suggested that the appropriate approach to fuzzy sets is to realize that the fuzziness simply is imposed by the fact of undetermined context. If we are surrounded by short people, then we will say a person of 5′6″ is tall; if we are meeting with the Watusi then such a person will be short. The idea, then, that a statement may have different truth values depending on the context suggests that there is implicit a whole series of mechanisms such as those that are being painfully developed in artificial intelligence approaches to the understanding of natural language (Schank and Colby, 1973). But such an idea takes us too far afield from the particular historical domain of discourse that we have set for ourselves in this paper, and so now we return to fuzzy sets, with the observation that one can clearly define a suitable functor T associated with 'fuzzing' (indeed, $QT = [0,1]^Q$), and that with this we may then define *fuzzy-set automata* to be those with dynamics $\delta : QT \to QT$, where T is the fuzzing functor.

With these three examples, we are ready to begin the development of our general theory. However, before we do so, it is worth making a couple of technical observations. Firstly, we may note that a continuous interval [0,1] may be replaced by any lattice, and that for technical reasons we shall usually want this to be a distributive lattice, and thus what is known as a semiring. In fact, Schützenberger (1962) has constructed a rich theory of automata over semirings so that not only are fuzzy sets a particular case of models already developed in multivalued and intuitionistic logic; but the study of fuzzy automata is a special case of Schützenberger's theory. Secondly, we note that Goguen (1967, 1969, to appear) has studied a category of fuzzy sets.

But all this is an aside, and it is time to return to the general study of dynamics of the form

$$\delta : QX \to QT$$

which provide the dynamics of what we call FUZZY MACHINES. [We hope that Professor Zadeh will forgive us for appropriating his word for this general setting – we use the term *fuzzy-set* machine to refer to his special case.] Our first observation is that $QX \to QT$ looks like a *generalization* of the case $QX \to Q$ which is obtained by taking T to be the identity functor. It would be far more appealing, aesthetically, if in fact we could take $QX \to QT$ to be a *special case*. But to do this we would have to consider a category $\mathcal{K}_\mathbf{T}$ whose objects are the same as those of the original category $\mathcal{K}$ but for which a morphism $A \to B$ is actually a $\mathcal{K}$-morphism $A \to BT$. In this case, a morphism $QX \to Q$, and thus a dynamics, in our new category $\mathcal{K}_\mathbf{T}$ would indeed be a morphism $QX \to QT$ in $\mathcal{K}$.

Recalling (Box 3) the need for *identities* and *composition* in defining a category, we can now develop a picture of what such a new category $\mathcal{K}_\mathbf{T}$ would look like. Our first requirement is that we can define identity morphisms for this category, and our choice for this is the morphism $Ae: A \to AT$ which tells us how to interpret pure elements as particular examples of fuzzy elements. For example, when $T=2^{(-)}$ we require $Ae: A \to 2^A$ to send an element a of set A to the singleton $\{a\}$ which is an element of the set 2^A of subsets of that set. Again, for $T=\mathbf{P}$, we require $Ae(a)$ to be the probability distribution on A for which a has probability 1. Given these identity morphisms, we can think of an ordinary morphism as a fuzzy morphism – namely we follow the morphism $A \to B$ with the 'fuzzing morphism' Be. Our second requirement in making $\mathcal{K}_T$ a category is a composition of fuzzy morphisms, so that we may compose $A \to BT$ with $B \to CT$ to obtain a morphism $A \to CT$ – in such a way that we have the usual axioms of a category for associativity of composition, and the existence of the identities which we require to be the 'fuzzing morphisms' Ae:

$$\mathcal{K}(A, BT) \times \mathcal{K}(B, CT) \to \mathcal{K}(A, CT) : (\alpha, \beta) \mapsto \beta \circ \alpha$$

which satisfies

$$(\gamma \circ \beta) \circ \alpha = \gamma \circ (\beta \circ \alpha)$$
$$\alpha \circ Ae = \alpha = \alpha \circ Be.$$

(We also require that $\beta \circ (Be \cdot f) = \beta \cdot f$ for $f: A \to B$, $\beta: B \to C$.) We call $\mathbf{T} = (T, e, \text{comp})$, and the category $\mathcal{K}_\mathbf{T}$ it induces, a **fuzzy category** over $\mathcal{K}$. (Adepts at category theory should note (MacLane, 1971; Manes, 1975a) that the notion of a fuzzy category is equivalent to the notion of a Kleisli category.)

Having introduced the idea of fuzzy category we find that there is a fly in the ointment, and it must be removed: We have been looking at $\mathcal{K}$-morphisms $QX \to QT$ and suggesting that the corresponding morphism from QX to Q in $\mathcal{K}_\mathbf{T}$ is a dynamics. But, unfortunately, so far we have only required X to be a functor on $\mathscr{K}$, not a functor on $\mathscr{K}_T$. This suggests, then, that we try to 'lift' the functor X on $\mathscr{K}$ to a functor $\tilde{X}$ on $\mathscr{K}_T$. Clearly, X and $\tilde{X}$ must act the same on objects. However, given a $\mathscr{K}$-morphism $A \to BT$, the action of X will yield a $\mathscr{K}$-morphism $AX \to BTX$, whereas $\tilde{X}$ will yield a $\mathscr{K}_T$-morphism $AX \rightharpoonup BX$, i.e., a $\mathscr{K}$-morphism from AX to BXT. We note that one way of reconciling this problem is simply to introduce for each object B a distinguished morphism

$$B\lambda : BTX \to BXT$$

Then define, for $g: A \to B = A \to BT$

$$g\tilde{X}: A\tilde{X} \to B\tilde{X} = AX \to BXT$$

to equal

$$AX \xrightarrow{gX} BTX \xrightarrow{B\lambda} BXT.$$

If X is to be a functor defined in this way, then λ must obey certain axioms which make it what a category theorist calls a **distributive law**. In fact, it can be verified that $\tilde{X}$ is a lift of X if and only if it is obtained from X by using a distributive law λ in this way. Thus, we may always denote $\tilde{X}$ by X_λ for the appropriate distributive law λ.

For example, in the case $X = - \times X_0$ and $T = 2^{(-)}$

$$Q\lambda : (2^Q) \times X_0 \to 2^{Q \times X_0} : (S, x) \mapsto \{(s, x) \mid s \in S\}$$

is the only distributive law.

More generally, replacing $2^{(-)}$ with *any* T: **Set** $\to$ **Set** gives rise to the distributive law

$$Q\lambda : QT \times X_0 \to (Q \times X_0)\, T : (P, x) \mapsto (in_x T)\, (P)$$

where

$$in_x : Q \to Q \times X_0 : q \mapsto (q, x).$$

Thus, there are many examples!

Once we have reached the stage of realizing that the proper setting for the study of nondeterministic automata is the category of some functor T using a functor X on $\mathscr{K}$ which can be lifted by a distributive law λ to a functor X_λ on $\mathscr{K}_T$ (Box 5) we can in fact show that many results holding for X are also available for X_λ. We can show that each X-dynamics 'is' an X_λ-dynamics, and we can show that each X_λ-dynamics may be 'simulated'

Fuzzy Machines

$$(X, T)\text{-Machines} \begin{cases} \tau: I \;\;\; \rightarrow QT \\ \delta: QX \rightarrow QT \\ \beta: Q \;\; \rightarrow Y \end{cases}$$

$X: \mathscr{X} \rightarrow \mathscr{X}$ is a functor and $\mathbf{T} = (T, e, \text{comp})$ is a fuzzy category for which there exists a distributive law $\lambda: TX \rightarrow XT$. Y is the carrier of a T-algebra. τ, δ and β are $\mathscr{X}$-morphisms.

Box 5

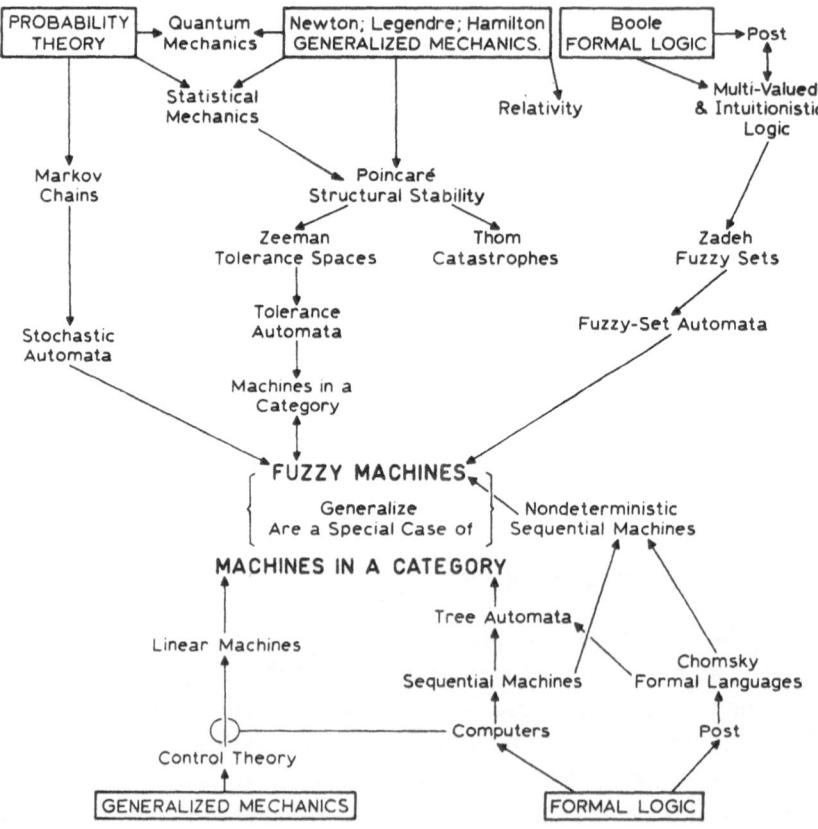

Fig. 5.

by an X-dynamics. Moreover, if we can do reachability theory for X, we can also do it for X_λ. If we can do observability theory for X we can also do it for X_λ if certain conditions concerning 'T-algebras' are met. Finally – and this is a technical comment whose content is clearly beyond the scope of this exposition – we may note that the proper setting for the theory of minimal realization for these fuzzy machines is the treatment of $(X, \mathbf{T})$-composite algebras.

Unfortunately, there is no space here to give the necessary background on category theory to expand upon any of these results, or the earlier results of Section I. However, we can summarize our discoveries quite succinctly. The idea of a morphism

$$\delta: QX \to Q$$

in a category $\mathscr{K}$ is the proper setting for the study of dynamics in a deterministic world. [We noted that the notions of left and right adjoint of a functor were crucial in studying reachability and observability, respectively, for such dynamics; as well as for approaching the theory of minimal realization.] What is perhaps most surprising is that dynamics in a fuzzy world is a *special case*, namely that in which the functor X is now an appropriate lifted functor X_λ, and the category in which the action takes place is a fuzzy category for some 'fuzzing functor' T. It is this 'surprise' that suggests that our general notion of a 'Machine in a Category' of Section I is indeed a proper setting for system theory: for one of the best tests of proper generality of a theory is that it is *robust* in the sense that it can admit apparent extensions as special cases, rather than requiring a proliferation of super- and subscripts for each new variation that arises. In conclusion, we synthesize our overview in the mandala of Figure 5.

BIBLIOGRAPHY

Arbib, M. A. and Manes, E. G.: 1974a, 'Machines in a Category: An Expository Introduction', *SIAM Review* **16**, 163–192.

Arbib, M. A. and Manes, E. G.: 1974b, 'Foundations of System Theory: Decomposable Machines', *Automatica* **10**, 285–302.

Arbib, M. A. and Manes, E. G.: 1975, *Arrows, Structures, and Functors: The Categorical Imperative*, Academic Press, New York.

Arbib, M. A. and Manes, E. G.: 1975, 'Fuzzy Machines in a Category', *Bull. Austral. Math. Soc.* **13**, 169–210.

Athans, M. and Falb, P. L.: 1966, *Optimal Control*, McGraw-Hill.

Bainbridge, E. S.: 1975, 'Addressed Machines and Duality', in E. G. Manes (ed.), *Category Theory Applied to Computation and Control, Lecture Notes in Computer Science* **25**, 93–98, Springer-Verlag, Heidelberg.

Bobrow, L. S. and Arbib, M. A.: 1974, *Discrete Mathematics*, Saunders, Philadelphia.

Ehrig, H., Kiermeier, K.-D., Kreowski, M.-J, and Kühnel, W.: 1974, *Universal Theory of Automata: A Categorical Approach*, Teubner.

Goguen, J. A.: 1967, '*L*-Fuzzy Sets', *J. Math. Anal. Appl.* **18**, 145–174.

Goguen, J. A.: 1969, 'The Logic of Inexact Concepts', *Synthese* **19**, 325–373.

Goguen, J. A.: 1972, 'Minimal Realization of Machines in Closed Categories', *Bull. Amer. Math. Soc.* **78**, 777–783.

Goguen, J. A.: 1973, 'Realization is Universal', *Math. Sys. Th.* **6**, 359–374.

Goguen, J. A.: 1974, 'Concept Representation in Natural and Artificial Languages: Axioms, Extensions and Applications for Fuzzy Sets', *Int. J. Man-Machine Studies* **6**, 513–561.

Goguen, J. A., Thatcher, J. W., Wagner, E. G., and Wright, J. B.: 1973, 'A Junction Between Computer Science and Category Theory, I: Basic Concepts and Examples (Part 1)', IBM Research Report RC 4526, T. J. Watson Research Center.

Kalman, R. E., Falb, P. L., and Arbib, M. A.: 1969, *Topics in Mathematical System Theory*, Mc-Graw Hill.

Lawvere, F. W. (ed.): 1972, *Toposes, Algebraic Geometry and Logic, Lecture Notes in Mathematics* **274**, Springer-Verlag.

MacLane, S.: 1971, *Categories for the Working Mathematician*, Springer-Verlag.

Manes, E. G.: 1975a, *Algebraic Theories*, Springer-Verlag.

Manes, E. G. (ed.): 1957b, *Category Theory Applied to Computation and Control, Proceedings of the First International Symposium, Lecture Notes in Computer Science* **25**, Springer-Verlag, Heidelberg.

Padulo, L. and Arbib, M. A.: 1974, *System Theory*, Saunders, Philadelphia.

Schank, R. C. and Colby, K. M. (eds.): 1973, *Computer Models of Thought and Language*, W. H. Freeman.

Schützenberger, M. P.: 1962, 'On a Theorem of R. Jungen', *Trans. Amer. Math. Soc.* **13**, 885–890.

Thom, R.: 1972, *Stabilité Structurelle et Morphogénèse*, W. A. Benjamin, Inc.

Zadeh, L.: 1965, 'Fuzzy Sets', *Inform. Control* **8**, 338–353.

CHAPTER 2

PARALLELISM, SLIDES, SCHEMAS, AND FRAMES*

MICHAEL A. ARBIB

In BT (Brain Theory), we study nets of simultaneously active neurons, and of interacting brain regions. In AI (Artificial Intelligence), we must structure programs for a serial computer. However, the development of a serial algorithm for a function does not preclude the existence of a more efficient parallel algorithm. For example, when adding two numbers, the propagation of the carry bit seems to force seriality. However, a look-ahead adder (see Hill and Peterson (1973) for a textbook treatment) can be built which uses parallelism based on 'carry look-ahead' to reduce addition time from the order of n (the length of the numbers) to the order of $\log n$ which, in fact, is the best possible (cf. Winograd's (1965)). Our task here is to examine the ways in which behavior is best expressed in structure, and consider the extent to which we can expect parallelism in that structure. Clearly, the 'precedence relations' of the real world – you must walk to the door before you go through it, for example – impose a high-level seriality on the flow of computation. However, within these high-level constraints, we shall see much room for parallel computation.

1. PARALLELISM

There is no question of the importance of parallelism in the early stages, at least, of visual processing. We see parallel extraction of 'bugness' in the frog retina, of contour and contrast information in mammals, and of other features in other animal visual systems. With their preprocessing cones, Riseman and Hanson have demonstrated the usefulness of parallel computation in layered structures as a first stage in scene analysis.

* This work was supported in part by NIH Grant No. 5 R01 NS09755-06 COM. Portions of the paper were presented at the Symposium on Parallel Processing in Artificial Intelligence held at New York University in January of 1975, and portions were presented at the George Hudson Symposium held at the State University of New York at Plattsburgh in April of 1975.

W. E. Hartnett (ed.), Systems: Approaches, Theories, Applications, 27–43.
*Copyright © 1977 by D. Reidel Publishing Company, Dordrecht-Holland.
All Rights Reserved.*

Didday's (1970) model of the frog tectum gives a low-level example of parallel decision-making – a network of 'sameness' and 'newness' elements acts in parallel upon its input array to extract the strongest (with exceptions analogous to those seen in frog behavior). In scene analysis, however, the recognition of, and the choice between, local features is not sufficient. We must pass from *local* to *semi-global* features – as when the local features of a door-frame define the enclosed area as a space through which we can walk. Dev (1975) has studied parallel networks for segmenting a scene into regions in a 'semantics-free' way, by having elements responding to a given feature in nearby locations excite other nearby detectors of that feature and inhibit detectors of other features. This results in the partition of the overall scene into regions in each of which only one type of feature detector is dominantly active. Riseman and Hanson use iterated computation up and down their cones to grow lines or regions of given texture. Burt's (1975) studies of networks which represent and support the movement of objects may give us clues as to how to use motion to aid region segmentation.

On the output side, we know that the brain uses activity in an array of motorneurons to control the populations of muscle fibres that constitute muscles. On the other hand, in AI the control of a stepping motor or rotary actuator does not seem to require inherent parallelism. However, there are other uses of parallelism. For example, Boylls (1975) modelled the cerebellum and its associated brain-stem nuclei as parameter-setting structures. We know that the basic algorithms for locomotion are in the spinal cord, but that the spinal animal does not 'shape' its steps properly. Stimulation of brainstem nuclei can increase muscular activity but – and this is the crucial point – Orlovsky (1972) found that in a walking animal, a muscle's activity is only increased during that phase of the step in which it should indeed be contracting. It is as if we have a motor control computer and a parameter-setting computer acting in parallel, but with the motor control computer only consulting the parameter setting when it is appropriate to do so.

Selfridge (1959) posited a character recognition system Pandemonium, which would behave as if there were a number of different (demons) sampling the input. Each demon was an expert in recognizing a particular classification and would yell out the strength of its conviction. An executive demon would then decree that the input belonged to the class of whichever demon it heard yelling the loudest. On the other hand, Kilmer, McCulloch and Blum (1969), in modelling the reticular formation, posited a system without executive control. Rather, each of an array of modules *sampled* the input and made a preliminary decision to the relative weights of different modes as being appropriate to the overall commitment of the organism. The modules were then coupled in a back-and-forth fashion so that eventually a majority of the modules would agree on the appropriate mode – at which stage the system would be committed to action. A reasonable analogy is a panel of physicians sharing symptoms and coming to a consensus about a diagnosis for a patient. (This suggests that social analogies may once again play an important role in brain theory.) Didday's (1970) model of the snapping behavior of a frog confronted with two flies, already mentioned in Section 1, posited a system of competitive interaction in the frog's tectum, which would lead in most cases to the suppression of all but one region of 'business' signalling, and result in the frog's snapping at one of the flies which caused the visual stimulation.

In attempting to place these studies in perspective, Montalvo (1975) observed that we could analyze all three models within a common framework, with the computational subsystems arrayed along two dimensions, one of competition and one of cooperation. In the Didday model, the cooperation dimension is 'degenerate' and the competition dimension is 'bug location'; in the Kilmer and McCulloch model competition is between 'modes' while cooperation is between 'modules'; and in the Dev model, competition is between 'disparities' and cooperation is along the 'space' dimension. The theme of competition and cooperation has thus emerged in three completely separate neural network models. It also plays a role when we look at the way in which an internal model of the world would operate.

Amongst the problems of a system receiving sensory input on the basis of which it must interact with the world are:

(i) **Segmentation**: To partition the input into 'segments' (not necessarily contiguous, nor confined to one modality) which define a single 'object' or other 'locus for possible interaction'. Dev (1975) provides a neural net model of segmentation processes in visual perception which shows how cooperation (consensus mechanisms) and competition of feature detectors can form part of a very low-level input-matching process. In the preprocessing cones of Hanson and Riseman (1975), more subtle routines – operating in parallel up and down several layers of preprocessors – are being developed for segmentation on non-primitive features, such as texture. Burt (1975) has modified the Dev model to support moving regions in response to moving inputs.

(ii) **Characterization**: The 'segments' are to be characterized in terms of 'programs for possible interaction'. [Stages (i) and (ii) are by no means sequential – some success at characterization may well aid the aggregation of distinct regions into a single 'segment'.]

(iii) **Relocation**: As the system moves, or as objects move in its world, the system must be able to easily update the internal representation of its world to take account of these changes. An important part of the updating is that an object which is moving uniformly is expected to continue doing so — Burt (1975) has neural net models which can support such moving representations, although the representations do not yet have the complex slide structure which we shall outline below.

(iv) **Tuneability**: With further 'exploration', or with changing goals, the system can tune its internal representation, either by increasing the level of resolution, or by emphasizing those features most relevant to the current goal structure.

(v) **Learning**: The system should not only change its representation on the basis of new input, but should be able to change the way in which that representation is constructed.

Didday and Arbib (1973) have built upon Didday's model of the frog tectum to suggest that human eye movements are controlled in part by computation in the superior colliculus akin to that taking place in the frog tectum, but with 'bugness' being replaced by a combination of peripheral signals, hypothesis signals, and mismatch signals — with the latter two classes of signals descending from the cortex.

Rosenfeld *et al.*'s (1975) study or region labelling also falls within this general theory of competition and cooperation (Arbib, 1975a), and the notion of cooperative computation finds support in a number of neurological studies such as those of Geschwind (1965), Luria (1973) and Nauta (1971) (see Arbib, 1975b).

2. SLIDES AND SCHEMAS

The **basic slide-box metaphor** (Arbib, 1972, p. 92) was intended as an antidote to a simple pattern recognition system in which the visual input pattern was to be classified as belonging to one of a small number of classes. Rather, the input pattern was to be analyzed as an array of familiar 'objects' — whether the object was a single object, such as a tree, or a composite object such as a row-of-trees — retrieving slides from a file, and arraying them in a slide-box to represent the current scene.

A critical notion was that *covering a portion of sensory input was to give access to appropriate programs for action* — though, since there are many objects, only some of the programs can 'take control' at any time, and some of them would be *planning* programs rather than programs for *overt* action. Thus some mechanisms must restrict which programs amongst this *redundancy of potential command* (to use McCulloch's phrase) will actually be implemented.

In an AI context, Minsky (1975) has developed a concept of *frame* which in some ways overlaps the above concept of slide, though with far more emphasis on linguistic and sociological aspects. We shall discuss frames in more detail in Section 3. (For more on computer understanding of language, see Schank and Colby (1973).) Intriguingly, the sociologist Erving Goffman (1974) has independently coined the term 'frame analysis' for analyzing the organization of experience − and his analysis has many points of overlap with Minsky's. Relevant ideas also occur in the approach to scene analysis espoused by Hanson and Riseman (1975).

In the rest of this section we outline an updated slide-box model for the organization of action-oriented memory for a perceiving system. A fuller account appears in Arbib (1975a, b). To avoid the overly pictorial connotations of the term "slide", I have adopted Piaget's term "schema", since much of the flavor of the tie-up between input-matching routines and action routines is contained in Piaget's notion of a schema (see Furth (1969) for an exposition). My concept is more formal and will hopefully (!) be shown to encompass all the more valuable aspects of Piaget's notion. I thus regard a schema henceforth as an array of programs to analyze a segment of the input to determine a possible course of action. As such, a schema must be locatable, tuneable, and linkable with other schemas. Thus, we have the following three components of a schema:

(i) **Input-Matching Routines**: A routine which succeeds to the extent that it covers a spatial region of multi-modal input (so that a cat schema can cover a furry region which emits meows, but not one that emits barks). This can be biased by non-sensory context inputs. The input-matching routines may include calls for confirming information (as in the eye-movement calls of Didday and Arbib (1975) − see also Szentágothai and Arbib (1974, pp. 335–339)). The level of success of these input-matching routines may be regarded as an 'activation' level of the schema, which increases as location and other parameters of the schema are adjusted to better fit the covered region. However, to complicate the story, the **resolution level** (i.e., the precision of this parameter match) required for activation to saturate may well depend on goal settings, or other non-sensory input to the schema. (In the *vision routines* of Hanson and Riseman (1975), the sensory input is purely visual − though contextual input is also used − and 'success' simply enables the assignment of a name to a region.)

(ii) **Action Routines**: The success of the input-matching routines in raising the activation level of a schema signals that certain actions have become appropriate for the system. Programs for some of these actions then form part of the schema. A crucial integrative property of schemas is that *increasing accuracy of parameter adjustment by the input-matching routines* *automatically adjusts parameters in the action routines in such a way that the action becomes more appropriate for the current environment and goal structure* (if the schema has been properly 'evolved').

(iii) **Competition and Cooperation Routines**: To date, we have talked of a schema as acting in isolation, attempting to raise its activation level by proper matching of input. But now we must realize that schemas are inter-connected. The operation of competition and cooperation routines helps determine which population of parameter-adjusted highly active schemas will constitute the current model of the environment. There still remains the problem of determining which of the action routines of these schemas are to operate – and another network of competition and cooperation routines will be involved in determining a compatible set of actions. (Since at any time, the organism is engaged in some activity, even if that be resting. Thus it is not so much a matter of choosing a course of action as it is of determining whether the time has come to change the course of action being pursued. The completion of an action may remove it from the competition. More interestingly, the execution of an action may provide new sensory input which de-activates the schema (or drastically changes the parameter settings) which enable the action – as when we bite into what appears to be a piece of fruit only to discover that it is made of wax.)

Notice that all these routines provide the *semantics* of a schema – what an object *means* to us comprises our knowledge of what we can do with the object and what relations it has with other objects, to the extent that our input-matching routines can capture the effects of these actions and relation-ships. In any case, we've come a long way from the original notion of a slide as being simply a colored transparency that approximates a region of the visual field.

For now, we shall assume that *all schemas may continually monitor their input pathways* (though different schemas have different input sets). In other words, the slide-file of the original metaphor becomes the total population of (relatively high-level) schemas of the present model; the slide-box of the original metaphor becomes the subpopulation of highly activated schemas of the present model. As in both Pandemonium and RETIC, we let each schema (i.e., mode-element) continually receive input. However – unlike both Pandemonium and the original slide-box metaphor – we shall for now try to do without a central executive overseeing the activation of schemas, and instead – in the spirit of RETIC – explore what can be achieved by the schemas themselves by virtue of their cooperation and competition routines. My methodological point is that it is not helpful to make *a priori* assumptions (whether to fit our preconceptions about neural net structure or about the utility of LISP programming) when setting up a framework of this generality. When we actually look at restricted systems which must be implemented in a brain or on a computer, then we can be more specific about the sets of executive and bookkeeping routines that seem necessary to augment the routines built into the schemas themselves.

In addition to schemas for objects, we may also have more abstract schemas (cf. the Hanson-Riseman *context routines*) such as one for 'winter'. Now at the change of seasons, the first fall of snow may be the signal for winter — so that we must posit the activity level of the 'snow-schema' providing an input to the 'winter-schema'. However, in the normal course of events, the organism *knows* that it is winter, and can use this **contextual information** to favor the hypothesis that a white expanse is snow rather than burnished sand, say, or moonlit water. It is this type of reciprocal activation (whether we regard it as an additional input, or as the action of a cooperation routine) that gives the system of schemas its **heterarchical** character. [Strictly defined, a 'heterarchy' is a system of rule by alien leaders. But in AI, stimulated by McCulloch (1949), it now denotes a structure in which a subsystem A may dominate a subsystem B at one time, and yet be dominated by B at some other time.] To the extent that the activation of a small population of schemas covers the activity of the feature-region schemas, to that extent can we say that the organism has perceived the scene.

3. FRAMES AND SCHEMAS

Given the complexity of physiological mechanisms that animals have evolved, we should expect the brain to be similarly sophisticated. Minsky (1975) thus posits a host of special-purpose mechanisms rather than a single *simple* mechanism. He introduces *frames* as his candidate for the unit underlying the effectiveness of common sense thought.

There seem to be three main reactions to Minsky's "Frames" paper:

(a) The "what a revelation" reaction of neophytes who had never before realized the importance of an internal representation of the world. Having confined their reading to a few recent theses and papers in AI, they were unaware of such contributions (to give but a limited sample) as Bartlett (1932), Craik (1943), Gregory (1969), MacKay (1955, 1963), Piaget (1954), Minsky (1961, 1965), and Young (1964).

(b) The "we've seen it all before" reaction. This comes in two flavors. (i) Some AI experts, having developed their own formalism for handling internal representations, dismiss Minsky's frames as a vague equivalent to their precise formulation. (ii) Other readers, familiar with the literature in (a), feel that Minsky was too cavalier in his brief reference to these earlier works, and object that many aspects of frames are ideas about internal representations with which they are long familiar.

(c) The "mature acceptance" reaction of workers in AI who have felt the need for a more general framework for the discussion of internal representations, and feel that recasting their work in the language of frames is a reasonable price to pay in moving toward such generality.

Having invested a moderate amount of effort in the slide-box metaphor and its extension to theory of schemas of Section 2, I must confess that my initial reaction was (b.ii). It is my feeling that Minsky's treatment of frames for scene analysis is inferior to mine in the sense that it seems too computer oriented rather than general enough to respond satisfactorily to the needs of brain theory. However, I recognize that Minsky's discussion of frames for language, understanding, and scenarios builds on recent work in computer understanding of natural language to handle aspects of internal representations for which my slides and schemata are too concrete. To this extent, I sympathize with reaction (c). The resolution of these apparently incompatible reactions is to distinguish more carefully than Minsky does between scene analysis and language understanding and discuss the extent to which they demand different styles of internal representation, the former being more slide/schema-like, the latter being more frame-like. The rest of this section discusses frame and schema for scene analysis; while Section 5 discusses pressures which require "protolinguistic" extensions to be made to the concept of a schema.

A frame includes information about

> How to use the frame
> What one can expect to happen next
> What to do if these expectations are not confirmed.

The top level represents things that are always true about the situation. Lower levels have *terminals* (slots) which are usually filled by 'subframes' which must meet certain conditions assigned at the terminal. In fact, a set of terminals may impose relations on their mutual assignments.

To handle the dynamics of a changing world Minsky posits a *system* of frames, transformations between which mirror the effects of important actions or changes in the world. Thus: a frame-system $\doteqdot$ a schema, and a transformation $\doteqdot$ updating schema parameters.

The theory of frames must handle change of percepts in the face of inconsistencies, errors, or new evidence; must explain imagery (cf. Bartlett (1932)) and must show how to exploit expectations. Minsky suggests that, to handle expectations, a frame's terminals are normally filled with default assignments − i.e., one cannot think of a ball without thinking of, say, a soccer ball of given size and colour. However Arbib (1975a) instead argues for a poset of schema-parameter specifications − and I would suggest that 'default' assignments may be very 'blurry' elements of the poset indeed, so that a ball may be little more specified than as requiring 'two hands to hold it', rather than being 'one-hand-holdable'. This is far from the level of precision that Minsky seems to suggest.

However if Minsky ascribes precision to *each frame,* he is insistent that the *frame-system* is far less precise than people believe their perception to be, citing the inability of all but the best draftsman to precisely render a variety of perspectives of a given object, and thus suggesting that a frame system comprises very few frames indeed. Minsky does not believe that our image changes as fast as does the scene. Rather, he posits that the illusion of continuity is due to the persistence of assignments to terminals common to different view frames − so that 'continuity' depends on the confirmation of expectations which in turn depends on rapid access to remembered knowledge about the visual world.

This analysis via discrete frames may be misguided. Stressing again our action-oriented view of perception, it is not the natural task of the system to draw pictures or to judge the accuracy of a drawing − though draftsmen can indeed master these skills. Rather, the task of the input matching routines is to activate schemas in a way which sets the parameters of action routines appropriate to the sensory input. Thus our shortcomings as draftsmen in no way imply a limitation of the accuracy of parameter-setting when we interact with real objects. We need not postulate an ability of the schemas to maintain very precise metrical relationships, for once foveal scanning has activated a schema, peripheral input − even though inadequate for object recognition − will suffice to maintain the appropriate place information when the object is no longer fixated.

Our viewpoint, then, is that an organism interacting with its world does *not* need a *complete* representation, but rather one that is easily *updated* as action progresses. [One of the greatest problems for the Shaky robot project was the lack of continuous visual input. In the same way, it is far easier to walk through a crowded room with our eyes open than it is to memorize the scene in sufficient detail to allow us to close our eyes and then walk to the door without bumping into anyone.] It is the range of actions in which the system will take part that will determine the appropriate level of detail − a map may have the metric relations all wrong so long as it reminds us to take the correct turning when we traverse the actual terrain − and so we see the effect of 'goal-setting schemas' upon the level of parameter-matching that will let a slide be sufficiently activated for its action-routines to be candidates for implementation.

In Section 1 we mentioned the work of Orlovsky as one dramatic instance of neural parameter-setting. It is thus natural for the brain theorist to posit a *tunable* system – once an object is identified, we simply update the parameters of the schema which represents it. By contrast (Figure 1) a frame represents an aspect of an object – with continuous changes in the input triggering discrete changes of frame. While this interchange of discrete structures may have some appeal for computer implementation, it appears unnatural for the brain, where varying frequencies of neural firing seem so well suited to represent continually changing quantities.

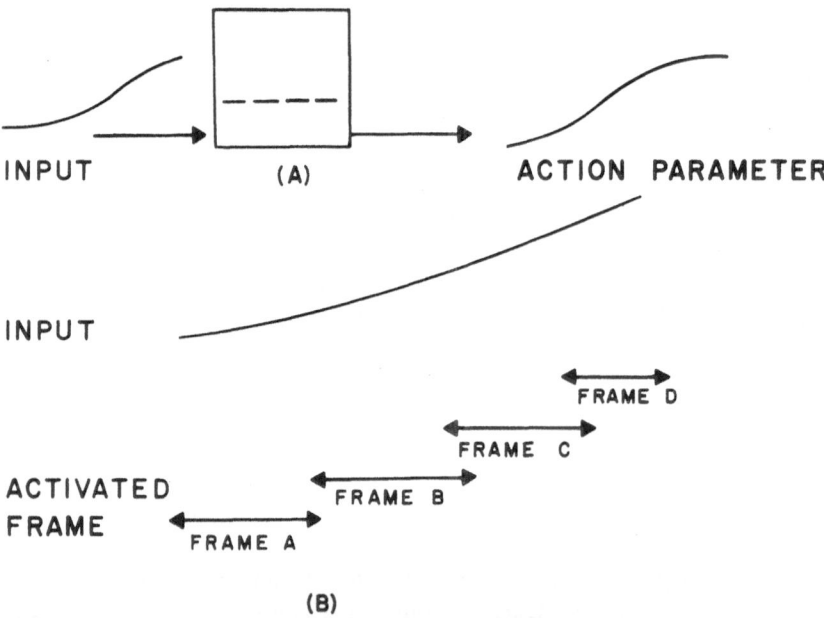

Fig. 1.   Schemas (A) versus Frames (B).

This leads us to the second crucial concept in the schema notion – it stresses the *generative nature* of the internal representations, which are built up as 'collages' of schemas. Minsky's framework seems to lack this concept – he talks of having one frame accessed at a time. For example, he posits one frame (in the form of a face-relation graph) for each *view* of a cube, and a different frame to represent the manipulative features of a cube. Once a

frame is proposed, a *matching* process tries to assign values to its terminals —
an information retrieval network provides a replacement frame if matching
fails. More specifically, Minsky posits the following process:

(1) Once a frame is evoked, it directs a test of its appropriateness, using
the current goal list to decide which terminals must match reality.

(2) It requests information to assign values to those terminals. (Failure
may cue evocation of an alternative frame.)

(3) Informed of a transformation (e.g., an impending motion) it would
transfer control to the appropriate frame of the frame system.

We thus see an emphasis on frames as "the complete internal representations of the current sensory input *per se*" rather than as "a component of a
representation which prepares the organism for interaction". The emphasis is
on a *serial process of frame selection* rather than the *parallel process of competition and cooperation posited in the schema model*; and little distinction
is made between changing a parameter and changing a hypothesis.

We may concede that a room provides a frame within which the recognition of doors and windows and their relationships is simplified. However, in
many cases, the framework is unimportant, and we directly recognize a
number of objects and come to grasp the situation in terms of their directly
perceived relationships. When we find an elephant sitting on our best chair,
we realize what is happening *in spite of* the framework of our expectations.
Our shock may be a measure of this discrepancy; but our understanding is a
strong argument for the "collage" approach as an important component of
internal representations: schemas trigger frames order schemas.... It is in
this more liberal sense of a frame as a *surrounding* that the concept seems
most useful; with that which fits in the frame, the schema, have somewhat
different properties. To the extent that we represent changes in objects,
parameter-setting in schemas seems appropriate. But the decomposition of
the world into objects imposes a discrete structuring onto a continuous
world, and the relationships between objects require a frame-like representation
closer to linguistics than to control theory. Perhaps a reasonable subgoal for
a theory of representations is to analyze the extent to which there is a
genuine distinction here and to what extent we are looking at poles of a con-
tinuum.

4. DEVELOPMENT

Turning to developmental questions, Minsky suggests that we compare
Piaget's *concrete operations* to the transformations between frames of a system (the tuning of parameters in a given schema); while the *formal stage*

might be characterized in terms of the ability to reason *about*, rather than simply to work *with*, those transformations. In computer terms, we might speak of the system developing a facility for writing 'commentaries' on its programs, or for reading its own programs. One might imagine that an 'abstract' of each schema comes to be developed with the schema itself, and that these are then available to 'higher-level' schemas. Is the language of frame-systems of schemas appropriate to the study of 'representations of representations'? I suspect that the answer is 'Yes'.

We have stressed the idea of a heterarchy of active schemas, all the way from 'line detectors' to abstract concepts like 'winter'. The input to a 'snow' schema is as much the activity of a context-schema like 'winter' as it is the output of some texture slide. Thus we begin to dissolve a strict interpretation of input-matching routines as matching *sensory* input, or of action routines as controlling *motor* output. Instead, we have a general system which may be involved in monitoring some schemas to better adjust the activity of others, rather than in sensorimotor correlation. But this is still consistent with our general definition of a schema. Thus schemas at one level can form "interschema operators" for schemas at another level. Such interschema operators would seem crucial in any theory of learning based on explicit hypothesis formation rather than mere synaptic adjustment.

Another Piagetian problem is addressed by Minsky in discussing *occlusion*. There might be *ad hoc* information about occlusion represented in a frame-system — so that we might have frames for each view of a chair being progressively occluded as it is slid under a table. Such frames probably do 'exist', but — more radically — one might posit a GLOBAL OCCLUSION SYSTEM which makes all perspective frames subsidiary to a central, common, space-frame system. The terminals of that subsystem would correspond to cells of a gross subjective space, whose transformations represent, once and for all, facts about which cells occlude others from different viewpoints. This certainly seems plausible for humans, and the work of Piaget and Inhelder suggest that complete coordination structures of this sort are not available to most children until they are at least ten years old. However, it would be naive to over-estimate the accuracy of this space-frame even in adults. Nepalese villagers never identify the two faces of a mountain if the faces can only be viewed from two different valleys separated by such ragged mountains that one must travel 100 miles to go from one vantage-point to the next; and many city-dwellers may drive a mile to get from one building to another down twisting one-way streets without realizing that they are within easy walking-distance of one another.

One of the really difficult problems of AI is to model the development of the space concept without building mechanisms into the original system which trivialize the whole problem. Consider, for example, how Ernst and Newell (1969) trivialized the *monkey-and-banana* problem when they implemented it in GPS. The monkey discovers that by moving a box and standing on that box, it can reach a banana suspended from the ceiling, and otherwise out of reach. But by providing their GPS simulation with 'vertical height' as an explicit difference, and by making 'climbing on the box' the only operator available to reduce that difference, Ernst and Newell excluded the only process that was genuinely of interest – the monkey's discovery that it could use the box as a tool. The question (being actively tackled by my colleague William Kilmer) is how to given the system so little information that the discovery is 'impressive', and yet enough information to enable the discovery to be made. The idea is to give the monkey a body-centered frame of reference, plus the ability to run, to bump into things, to reach and grasp, and to climb. It can then discover, as a result of several 'crashes', that boxes are moveable, and learn – through *play* – how to move them in a deliberate way. Again, it may learn that climbing on boxes gets it further from the ground. It is a challenge to our ability not to build too much in to make it a *separate* discovery – but one not so surprising in a body-centered, as distinct from Euclidean, framework – that this climbing also gets it closer to the ceiling. Then, and only then, is it ready to solve the problem of reaching the bananas hanging from the ceiling. Note here the importance of *play* in providing mechanisms for later problem-solving.

5. MORE ON PARALLELISM

Didday and Arbib (1975) studied eye movements and visual perception with a hybrid model in which a 'slide-box' cortex interacted with a midbrain system adapted from Didday's neural net model of the frog tectum. The constraints imposed on cortex by interfacing it with a midbrain system so structured will suggest ways to move towards a more neural model of cortex.

We perceive a scene via a *series* of visual fixations, required to bring successive regions before the fovea; although the above model suggests that the computation required to determine the eye movement and process the input is parallel. Is there any reason why the whole retina does not have foveal acuity, allowing the whole process of analyzing a scene to be accomplished in parallel upon a single fixation? The frog's 'bug-detectors' operate in parallel without eye movements. But a 'SUPERFROG' should not simply snap at flies, but should also learn about new objects in its world. If the whole recognition machinery were iterated over the whole visual field, there would be

the problem of communicating information learnt about an object which has
appeared in one part of the visual field to the machinery which would handle
its appearance in each other region of the visual field. It may well be [a
careful mathematical analysis is called for] that it is computationally effective to use, e.g., eye movements to route visual input to a standard processor
than it is to route learned information to a host of parallel processors for the
recognition of a given class of objects.

Seriality, then, is imposed on visual perception by the sequence of eye
movements in visual perception; and we may note, too, the seriality of
speech, as if each word were trying to direct a fixation of the attention of
the listener. Arbib (1975a) discusses the notion of building a 'superschema'
from a repeated scene, thus providing larger units of representational activity.
This accords well with Minsky's view that rapid selection of large substructures
— to provide the context in which selection of 'subframes' takes place — will
speed perception and thought. Though we disagree with Minsky's insistence
on purely serial processing, the time has come to face up to the fact that
there must be limits to the parallelism which a slide-box or collage of schema
can handle.

We have talked of a schema for each concept, be it 'tree', 'winter', 'differential calculus' or 'ontological commitment'; and we've offered the set of
currently active schemas, together with their parameter settings, as providing
the internal representation of the current state —both internal and external
 - of the organism's perceived world. But how do we handle multiple objects?
Do we imagine that we have several copies of each schema, and that the competition routines linking them are sufficiently strong that only n can be
activated when n of the objects that schema represents are in the environment?
Is it, then, that for each object, there is an upper bound on the number of instances we can apprehend — 17 trees, and no more?! Certainly, the discussion
of 'superfrog' makes this option unappealing — with multiple three schemas,
how do we share the adaptive changes in one with the other 16?

Perhaps — and I do not yet know how to phrase this in neural terms, but
must use a simple-minded computerese — we should imagine a single copy of
each schema, but posit that it can accommodate several pointers, with appropriate settings of location and other parameter settings for each object which
provides the 'source' of such a pointer. Calling on the folklore of anthropology
— stories of primitive tribes that count 'one, two, three, many' — it seems
reasonable to posit an upper bound of three, say, to the number of pointers
which can bear detailed parametric information. After this, 'lumped' form of
description — a 'row-of-trees', say - may be the most explicit representation
one can handle without linguistic intervention. Thus a human can count 17

trees, and remember that there are indeed seventeen — but this is a more abstract form of representation than the process of parameter tuning by input matching routines. [It is intriguing to speculate on the extent to which these two types of parameter settings — linguistic and non-linguistic — may be localized in the left and right hemispheres of humans; and to explore the role of the *corpus callosum* in integrating these two types of information. It may be — returning to our discussion of Piaget — that the distinction between the hemispheres is related to the distinction between formal and concrete operations. However, this is probably too crude a division of labour.]

In any case, we see that schema must be able to 'quell' several regions, rather than simply one; and that a significant step in evolution — to avoid undue demands on parallelism — was the ability to move from 'quelling' by precise parameter adjustment by the input-matching routines to 'quelling' by a more abstract, proto-linguistic, representation which could, for example, simply note the number and approximate disposition of an array of similar objects. [Incidentally, I would suggest that this multitude of simultaneous activity is what makes perception 'richer' than imagining — we not only have 'tree' schemas active during perception, but a rich array of texture and other 'low-level' schemas, too. Note, too, that dreaming is a natural facet of the schema model: schemas can activate one another in complex activity patterns even with the reduced sensory input that characterizes sleep.]

The transition to proto-linguistic parameter is one way of overloading the capacity of any one schema. Another approach is to make 'copies' of the schema tuned to different instances of a given type — as when we differentiate the schema for 'man' into one for each of the men with which we are at all acquainted, from a very sketchy schema for a public figure seen occasionally in the newspaper, to a schema, far richer than the generic schema, for a very close friend.

What does this say for the concept of identity? At one level, we may say an object, or a low-level pattern of schema activity, is more-or-less identical to another to the extent that — in a given context — it yields the same pattern of action or high-level activity. However, it is one thing for the organism to *behave* as if the two are identical; it is another thing for it to be *aware* of this identity. Presumably, this 'awareness' requires the 'schema-abstracts' posited in the discussion of Piaget's formal operations, together with mechanisms to compared 'abstracts' activated by the two patterns.

In his article, 'The Architecture of Complexity', H. A. Simon argues for hierarchical structuring of complex systems, suggesting that evolution can more effectively act upon a system made up of functionally well-defined subsystems. Minsky argued similarly, going from the need to debug knowledge systems (cf. Winston's program for learning structural descriptions by debugging a preliminary description by using examples and near-misses) to the need for structured programs. However, he seems mistaken when he suggests (admittedly in the role of 'Devil's Advocate') that such a structured program must be serial — any more than the evolution of organs should require hearts, lungs, and liver to be time-shared!

Between the admittedly parallel input and output structures lies the region of 'congnitive computation', and Minsky claims that this is inherently serial. In fact, we know that different regions of the brain communicate during cognition — one can get auditory tuning curves from visual cortex neurons — as if each region were trying to model the world on the basis of its own primary data, and yet keep the model consistent with information about the activity of other model builders (so that the 'internal model' in our heads is not a unitary construct, but is a population of models, agreeing in crude outline, but differing in type and depth of detail). However, it is striking that while this array of parallel subsystems lets us, for example, recognize with alacrity a sought-for object in a complex scene, our billions of neurons may take several seconds to add a pair of 4-digit numbers. This suggests that the natural parallelism expressed in our brains by the multiplicity of anatomically distinct regions may have evolved to suit us for a primitive hunting existence, but be little adapted for the linguistic and cultural 'computations' which mankind has evolved since our brains achieved their present form. This may impose a semblance of seriality on many such 'computations'. However — recall the look-ahead adder — this does not preclude the incorporation of far greater parallelism in a computational structure specifically designed for socio-linguistic behavior.

In summary, our concern in both AI and BT is with the mediation of complex behavior by appropriate structures. In each case, questions of efficiency, evolution, learning, and 'debuggability' will enter, and it can be expected that the temporally serial execution of a variety of processes operating in parallel will provide the proper setting for their analysis.

REFERENCES

Arbib, M. A.: 1972, *The Metaphorical Brain*, Wiley-Interscience, New York.
Arbib, M. A.: 1975a, 'Segmentation, Schemas, and Cooperative Computation', in G. Levin (ed.), *MAA Studies in Mathematics*, Biomathematics.
Arbib, M. A.: 1975b, 'Artificial Intelligence and Brain Theory: Unities and Diversities', *Ann. Biomed. Eng.*
Bartlett, F. C.: 1932, *Remembering*, Cambridge University Press.
Boylls, C. C.: 1975, 'The Function of the Cerebellum and its Related Nuclei as Embedded in a General Paradigm for Motor Control', Technical Report, Computer and Information Science, University of Massachusetts at Amherst.

Burt, P.: 1975, 'Computer Simulation of a Dynamic Visual Perception Model', *Int J. Man-Machine Studies.*

Craik, K. J. W.: 1943, *The Nature of Explanation*, Cambridge University Press.

Dev, P.: 1975, 'Segmentation Processes in Visual Perception: A Cooperative Neural Model', *Int. J. Man-Machine Studies.*

Didday, R. L.: 1970, 'The Simulation and Modelling of Distributed Information Processing in the Frog Visual System', Ph.D. Thesis, Stanford Univ.

Didday, R. L. and Arbib, M. A.: 1975, 'Eye Movements and Visual Perception: a "Two Visual System" Model', *Int. J. Man-Machine Studies.*

Ernst, G. W. and Newell, A. W.: 1969, GPS: *A Case Study in Generality and Problem Solving*, Academic Press.

Furth, H. G.: 1969, *Piaget and Knowledge*, Prentice-Hall.

Geschwind, N.: 1965, 'Disconnexion Syndromes in Animals and Man', *Brain* 88 Part I, 237–294, Part II, 585–644.

Goffman, E.: 1974, *Frame Analysis: An Essay on the Organization of Experience*, Harper Colophon Books.

Gregory, R. L.: 1969, 'On How so Little Information Controls so Much Behavior', in C. H. Waddington, (ed.), *Towards a Theoretical Biology 2 Sketches*, Edinburgh University Press.

Hanson, A. L. and Riseman, E. M.: 1975, 'The Design of a Semantically-Directed Vision Processor', COINS Technical Report 75C-1, University of Massachusetts at Amherst.

Hill, F. J. and Peterson, G. R.: 1973, *Digital Systems: Hardware Organization and Design*, Wiley, New York.

Kilmer, W. L., McCulloch, W. S. and Blum, J.: 1969, 'A Model of the Vertebrate Central Command System', *Int. J. Man-Machine Studies* 1, 279–309.

Luria, A. R.: 1973, *The Working Brain*, Penguin Books.

MacKay, D. M.: 1955, 'The Epistemological Problem for Automaton', in *Automata Studies*, C. E. Shannon and J. McCarthy (eds.), Princeton University Press.

Mackay, D. M.: 1963, Internal Representation of the External World, *AGARD Symposium of Natural and Artificial Logic Processors*, Athens, Mimeographed, 14 pages.

Minsky, M. L.: 1961, Steps Towards Artificial Intelligence, *Proc. IRE* 49, 8–30.

Minsky, M. L.: 1965, Matter, Mind and Models, *Information Processing 1965, Proc. IFIP Congress* 1, 45–49.

Minsky, M. L.: 1975, 'A Framework for Representing Knowledge', in P. H. Winston (ed.), *The Psychology of Computer Vision*, New York, McGraw-Hill, pp. 211–277.

Montalvo, F. S.: 1975, 'Consensus versus Competition in Neural Networks', *Int. J. Man-Machine Studies* 7, 333–346.

Nauta, W. J. H.: 1971, 'The Problem of the Frontal Lobe: A Reinterpretation', *J. Psychiat. Res* 8, 167–187.

Orlovsky, G. N.: 1972, 'The Effect of Different Descending Systems of Flexor and Extensor Activity During Locomotion', *Brain Research* 40, 359–372.

Piaget, J.: 1954, 'The Construction of Reality in the Child', Basic Books.

Rosenfeld, A., Hummel, R. A. and Zucker, S. W.: 'Scene Labelling by Relaxation Operators', TR. 379, Computer Science Center, University of Maryland.

Schank, R. C. and Colby, K. M. (eds.): 1973, *Computer Models of Thought and Language*, W. H. Freeman.

Selfridge, O. L.: 1959, 'Pandemonium: A Paradigm for Learning', *Mechanization of Thought Processes*, London: H.M.S.O., 513–526.

Szentágothai, J. and Arbib, M. A.: 1974, Oct. *Conceptual Models of Neural Organization, Neurosciences Research Program Bulletin* 12: No. 3.

Winograd, S.: 1965, 'On the Time Required to Perform Addition', *J. Assoc. Comp. Mach.* 12, 235–243.

Young, J. Z.: 1964, *A Model of the Brain*, Oxford University Press.

CHAPTER 3

THE FUNDAMENTAL DUALITY OF SYSTEM THEORY

E. S. BAINBRIDGE

1. INTRODUCTION

A system may be specified in either of two ways. In the first, which we shall
call a *state description*, sets of abstract inputs, outputs and states are given,
together with the action of the inputs on the states and the assignments of
outputs to states. In the second, which we shall call a *coordinate description*,
certain input, output and state variables are given, together with a system
of dynamical equations describing the relations among the variables as functions of time. Modern mathematical system theory is formulated in terms of
state descriptions, whereas the classical formulation is typically a coordinate
description, for example a system of differential equations. Current preference
for the state description apparently arises from the presumption that this is
the abstract or invariant description of the system. Each coordinate description determines just one state description, but that state description can be
represented by many distinct coordinate descriptions. To illustrate the distinction, as usually understood, between state descriptions and coordinate
descriptions, and the relationship between them, we consider an example.
Throughout this paper we use discrete systems as typical of the general
case.

Definition 1.1. Relative to fixed sets Σ of *inputs* and V of *outputs*, an
automaton $A = (Q, i, \beta)$ consists of a set Q of *states*, a distinguished *initial
state* $i \in Q$, an *output function* $\beta: Q \to V$, and a *state transition function*

$$((q, \sigma) \mapsto q\sigma): Q \times \Sigma \to Q.$$

Denote by Σ^* the free monoid generated by Σ, writing 1 for the unit element
of Σ^*. Extend the state transition function to a right action of the monoid
Σ^* on Q

$$((q, x) \mapsto qx): Q \times \Sigma^* \to Q$$

W. E. Hartnett (ed.), Systems: Approaches, Theories, Applications, 45–61.
Copyright © 1977 by D. Reidel Publishing Company, Dordrecht-Holland.
All Rights Reserved.

defined by $q1 = q, q(\sigma x) = (q\sigma)x$. The *behaviour* of an automaton $A = (i, Q, \beta)$ is the function

$$(x \mapsto \beta(ix)): \Sigma^* \to V. \quad //$$

Automata are, by this definition, to be specified by state descriptions. Other definitions have been used; for example, automata have been defined as formal circuits, that is, by coordinate descriptions of the type exemplified below. We shall not formally define a circuit, since we shall later argue for a somewhat different formal explication of this type of coordinate description.

Example 1.2. A Coordinate Description.

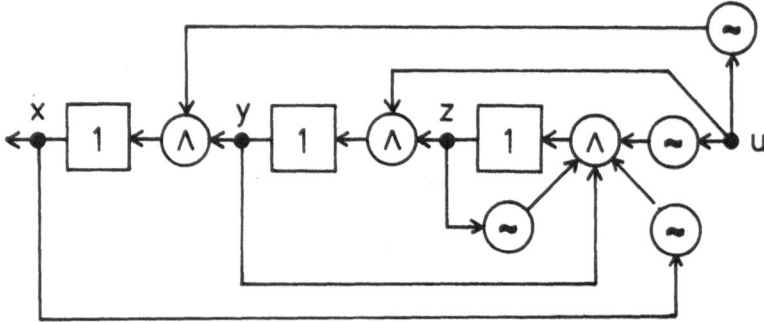

The conventions of this *circuit diagram* are as follows. Each box represents a unit delay element which holds the (binary) value of some state variable and bears a label giving the initial value of that variable. The output variable appears as the label of a node having an outgoing arrow with no target. The circles represent components which instantaneously compute the indicated Boolean functions. Evidently such a diagram is equivalent to the following equations, together with a specification that x is the output.

$$x(0) = 1, \; y(0) = 1, \; z(0) = 1 \tag{1.2.1}$$

$$\left. \begin{array}{l} x(t+1) = y(t) \wedge \sim u(t) \\ y(t+1) = z(t) \wedge u(t) \\ z(t+1) = \sim x(t) \wedge y(t) \wedge \sim z(t) \wedge \sim u(t). \end{array} \right\} \tag{1.2.2}$$

The relationship between state descriptions and coordinate descriptions may be illustrated with this example. Take as states all binary triples xyz representing assignments of values to the state variables. The state 111 is initial, by Equation (1.2.1). The output function β is given by

$$\beta(xyz) = x$$

since x is the output variable. The state transition function is specified by Equations (1.2.2) as

$$(xyz)u = (y \wedge \sim u)(z \wedge u)(\sim x \wedge y \wedge \sim z \wedge \sim u).$$

Conversely, given the eight state automaton just described, one can choose any encoding of the states as distinct binary triples and, by a well known construction, obtain a circuit which realizes this same automaton. Apparently the state description represents the abstract system, and the various coordinate descriptions are mere codings of the underlying abstract automaton. //

All of what we have said is familiar ground, indeed the ground on which the foundations of modern system theory rest. We claim, however, that system theory can be built on a quite different foundation. In this alternative system theory a suitable notion of abstract coordinate description replaces the abstract state description as the definition of system. The development of this theory can be carried forward entirely without reference to the concept of state. It is not appropriate to say that these system theories are equivalent; rather, they are *dual* in a sense which will be made precise.

This alternative system theory is developed for the special, but typical, case of automata in the next section.

2. NETWORKS

In the coordinate description 1.2., the variables y and z are so related that the value of y after input 1 is always equal to the value of z before input 1. This follows from the equation

$$y(t+1) = z(t) \wedge u(t).$$

If we view the variables x, y, z as memory locations in the circuit, we might say that input 1 causes the information stored in location z to be transferred to location y. For input 0, however, the next value of y is always 0. Since by Equation (1.2.1) the initial values of x, y, z are 1, it cannot be said of any location $r \in \{x, y, z\}$ that input 0 causes the information in location r to be transferred to y. We propose to formalize this notion of information flow between variables and consider a system theory in which information flow, rather than state transition, is basic.

Example 2.1. An Abstract Coordinate Description.

$$x(0) = 1, \ y(0) = 1, \ z(0) = 1, \ v(0) = 0, \ w(0) = 0 \qquad (2.1.1)$$

$$\left.\begin{array}{rcl}
x(t+1) & = & y(t) \wedge \sim u(t) \\
y(t+1) & = & z(t) \wedge u(t) \\
z(t+1) & = & w(t) \wedge \sim u(t) \\
v(t+1) & = & v(t) \\
w(t+1) & = & z(t) \wedge u(t).
\end{array}\right\} \qquad (2.1.2)$$

These equations may be viewed as a reformulation of Example 1.2 in the following way. The additional equations

$$w(t) = \sim x(t) \wedge y(t) \wedge \sim z(t) \qquad (2.1.3)$$
$$v(t) = 0 \qquad (2.1.4)$$

are consequences of Equations (2.1.1) and (2.1.2), as one can easily show by induction on t. The equations of this example for the variables x, y, z are therefore equivalent to the equations of Example 1.2.

Moreover, for each variable r and each input σ, there is a variable r' such that for all $t \geqslant 0$

$$u(t) = \sigma \quad \text{implies} \quad r(t+1) = r'(t).$$

This follows from the form of the equations; for example, for $r = y$, if $u(t) = 1$ then

$$y(t+1) = z(t) \wedge 1 = z(t)$$

and if $u(t) = 0$, then

$$y(t+1) = z(t) \wedge 0 = 0 = v(t)$$

by Equation (2.1.4).

Thus the Equations (2.1.1) and (2.1.2), together with the specification of x as output, may be summarized in the following *information flow diagram*, explained below.

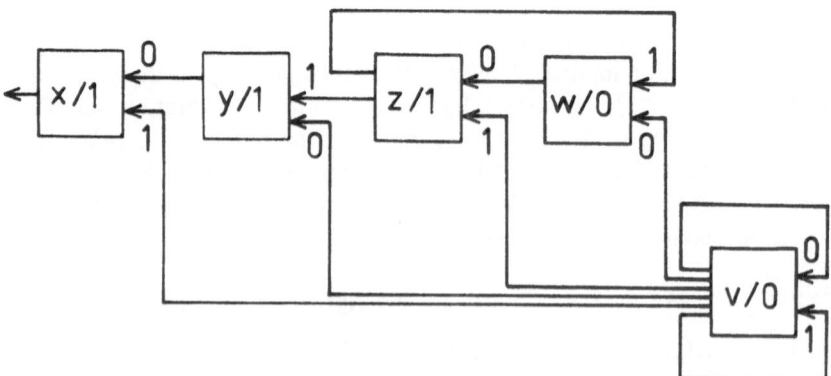

The boxes are labelled by the variables, together with the values given in Equations (2.1.1). The output coordinate is distinguished by an outgoing arrow with no target. The flow of information is determined by the input as shown by the arrows and their labels.

We can explain the operation of an information flow diagram by introducing the state of the system, and will do so now as a guide to intuition, even though we shall later explicitly forgo state-oriented language. Initially each box contains the value given by its label. On input σ, the contents of each box is transmitted along all outgoing wires labelled σ. The new contents of each box is the value arriving on the unique incoming wire labelled σ. The output is the contents of box x. What we have just specified is the state description of the system described by the information flow diagram. //

We shall now begin a formal presentation of a system theory which takes information flow as its primitive concept.

Definition 2.2. Relative to fixed sets Σ of *inputs* and V of *outputs*, a *network* $N = [R, j, \alpha]$ consists of a set R of *variables*, a distinguished *output variable* $j \in R$, a *values function* $\alpha: R \to V$, and an *information flow function*

$$((\sigma, r) \mapsto \sigma r): \Sigma \times R \to R.$$

Extend this to a left action of the monoid Σ^* on R

$$((x, r) \mapsto xr): \Sigma^* \times R \to R$$

defined by $1r = r$, $(x\sigma)r = x(\sigma r)$. The *behaviour* of a network $N = [R, j, \alpha]$ is the function

$$(x \mapsto \alpha(xj)): \Sigma^* \to V. //$$

Although a network is essentially the same formal object as a finite automaton, the behaviour of this object as a network is not its behaviour as an automaton. We motivate this by introducing a state-free language for the operation of a network.

A network should be visualized by means of its information flow diagram, whose operation is described informally by saying that on input σ information flows along arrows labelled σ. The concept of information flow is not to be further analysed. The formal counterpart of information flow along arrows labelled σ is a relation σ from R to R; namely, that exhibited in the diagram. By construction, the reciprocal relation σ^{-1} is a function; that is, for each $r \in R$ there is a unique $r' \in R$ such that $r'\sigma r$. It follows that after a sequence $\sigma_1 \sigma_2 \ldots \sigma_n$ of inputs, there is, for each $r \in R$, a unique $r' \in R$ such that $r'(\sigma_1 \sigma_2 \ldots \sigma_n)r$. In particular, for each input sequence, there is a unique

$r \in R$ from which information flows to the output coordinate j. Thus we may view a network as an information retrieval system in which each input sequence *retrieves* (that is, causes to flow to j) the information stored at some $r \in R$. The value $\alpha(r) \in V$ formalizes the expression "information stored at r". In the formal Definition 2.2, σr denotes the unique coordinate r' such that $r' \sigma r$. Thus for $x \in \Sigma^*$, xr is the unique element of R from which, information flows to r on input x. From the above discussion, the value retrieved by input $x \in \Sigma^*$ is the value stored at xj; that is, the value $\alpha(xj)$. This motivates the definition of network behaviour.

Remark 2.3. For a network $N = [R, j, \alpha]$ we may associate to each input sequence $\sigma_0 \sigma_1 \ldots \sigma_n$ the sequence $v_1 v_2 \ldots v_{n+1}$ of values retrieved. These values are the values of the behaviour function, that is

$$v_{k+1} = \alpha(\sigma_0 \sigma_1 \ldots \sigma_k j).$$

For an automaton $A = (Q, i, \beta)$ we may associate to an input $\sigma_0 \sigma_1 \ldots \sigma_n$ the sequence of output values $u_1 u_2 \ldots u_{n+1}$ where

$$u_{k+1} = \beta(i\sigma_0 \sigma_1 \ldots \sigma_k).$$

It may not be immediately obvious that these situations are *not* analogous. For the automaton, the underlying sequence of states $q_{k+1} = i\sigma_0 \sigma_1 \ldots \sigma_k$ has the property that q_k and σ_k determine q_{k+1}. For the network, the underlying sequence of variables $r_{k+1} = \sigma_0 \sigma_1 \ldots \sigma_k j$ need not have this property, as the following example shows. In the information flow diagram of Example 2.1., input 010 retrieves the sequence of variables yvw and the input 110 retrieves vvv. In the first case we have $r_2 = v$, $\sigma_2 = 0$ and likewise in the second case; but in the first case $r_3 = w$ and in the second case $r_3 = v$. //

Example 2.4. The information flow diagram shows how the system can be realized as an interconnection of block components. For example, by considering the information flow diagram of 2.1 we obtain a series decomposition which in conventional block diagram notation would appear as below, where the state of each component system is given by the values of the variables labelling it. As we shall see later, the decomposition theory which this yields corresponds precisely to Hartmanis–Stearns (1966) decomposition of the associated state description, which is considerably more complicated since it deals with equivalence relations rather than subsets. //

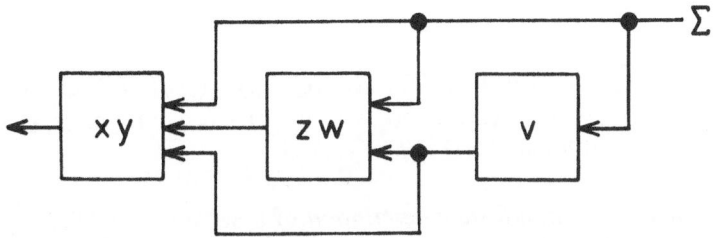

Remark 2.5. One can show, without introducing the notion of state (not even in a disguised form) that for every function $f\colon \Sigma^* \to V$ there is a canonical minimal network having behaviour f. This is an example of the type of result which can be obtained in a system theory based on abstract coordinate descriptions. The proof is formally similar to the proof of the analogous result for state descriptions. In this section we have argued that system theory *could* be founded on information flow rather than on state transition. Some things would be harder (2.3), and some things would be easier (2.4); but one would have a self-contained system theory comparable to that based on state descriptions. In the next section we propose that neither system description should have special status.   //

3. DUALITY

The duality principle for systems which we consider in this section was first enunciated by Kalman (1960) for the special case of linear systems. The surprising fact that Kalman's duality does not rest on the duality of finite dimensional vector spaces, but can be formulated for more general systems was first observed by Wymore (1967). Other authors (Arbib and Zeiger (1969); Bainbridge (1972), (1975); Arbib and Manes) have considered versions of this duality.

The contribution of this paper to the understanding of this duality consists of:

(i) The interpretation of the duality as a *representation theory* relating state descriptions and coordinate descriptions.

(ii) The interpretation of the minimizing of a reachable system as a double dualization.

(iii) The unification of the state and coordinate viewpoints via the notion of a *coordinatized system*.

As before we consider the special case of automata, the generalization to other types of system being quite clear.

3.1. *Representations*

Definition 3.1.1. A *homomorphism* φ: $(Q, i, \beta) \to (Q', i', \beta')$ of automata is a function $\varphi: Q \to Q'$ such that $\varphi(i) = i'$; $\varphi(q\sigma) = (\varphi q)\sigma$ for all $q \in Q$, $\sigma \in \Sigma$; and $\beta(q) = \beta'(\varphi q)$ for all $q \in Q$. //

Definition 3.1.2. The *full state description* of a network $N = [R, j, \alpha]$ is the automaton $A(N) = (V^R, \alpha, \tilde{j})$, where $\tilde{j}: V^R \to V$ is given by $\tilde{j}(f) = f(j)$, and the state transition function

$$((f, \sigma) \mapsto f\sigma): V^R \times \Sigma \to V^R$$

is given by $(f\sigma)(r) \stackrel{\frown}{=} f(\sigma r)$. //

The full state description formalizes the passage from coordinate descriptions to state descriptions. Behaviour is an invariant of this construction, as follows.

Proposition 3.1.3. The behaviour of $A(N)$ is that of N.
Proof. The behaviour function of $A(N)$ is $x \mapsto \tilde{j}(\alpha x)$. However, $\tilde{j}(\alpha x) = (\alpha x)(j) = \alpha(xj)$ by induction on $x \in \Sigma^*$. But $x \mapsto \alpha(xj)$ is the behaviour function of N. //

Definition 3.1.4. A *representation* of an automaton A by a network N is a homomorphism $A \to A(N)$. An injective representation is called a *realization*. //

The point of view implicit by these definitions is the usual one that state descriptions are basic and that coordinate descriptions should be understood by reduction to state descriptions. That such a viewpoint has no special status is demonstrated by existence of a dual reduction.

Definition 3.1.1*. A *homomorphism* ψ: $[R, j, \alpha] \to [R', j', \alpha']$ of networks is a function $\psi: R \to R'$ such that $\psi(j) = j'$; $\psi(\sigma r) = \sigma(\psi r)$ for all $r \in R$, $\sigma \in \Sigma$; and $\alpha r = \alpha'(\psi r)$ for all $r \in R$. //

Definition 3.1.2*. The *full coordinate description* of an automaton $A = (Q, i, \beta)$ is the network $N(A) = [V^Q, \beta, \tilde{i}]$ where $\tilde{i}: V^Q \to V$ is given by $\tilde{i}(g) = g(i)$, and the information flow function

$$((\sigma, g) \mapsto \sigma g): \Sigma \times V^Q \to \Sigma$$

given is by $(\sigma g)(q) = g(q\sigma)$. //

The full coordinate description formalizes the passage from state descriptions to coordinate descriptions. The network $N(A)$ represents the flow of information among all V-valued attributes of Q. By a dual argument, behaviour is an invariant of the construction $A \mapsto N(A)$.

Proposition 3.1.3*. The behaviour of $N(A)$ is that of A. //

Definition 3.1.4*. A *representation* of a network N by an automaton A is a homomorphism $N \to N(A)$. An injective representation is called a *realization*. //

A network represents an abstract pattern of information flow. To represent the network by an automaton is to represent the abstract variables as V-valued attributes on some set, and the abstract information flow as the information flow induced by transitions on that set. The use of the term 'representation' in 3.1.4 and 3.1.4* should be compared to its other uses in mathematics; for example, the representation of a Boolean algebra by the subsets of some set.

Definition 3.1.5. Let $A = (Q, i, \beta), N = [R, j, \alpha]$. If $\varphi: A \to A(N)$ is a representation, the *dual representation* $\tilde{\varphi}: N \to N(A)$ is given by the function $\tilde{\varphi}: R \to V^Q$ defined by

$$(\tilde{\varphi}r)(q) = (\varphi q)(r). \quad //$$

Proposition 3.1.6. The dual $\tilde{\varphi}: N \to N(A)$ of a representation $\varphi: A \to A(N)$ is a representation.

Proof: Let $N = [R, j, \alpha]$, $A = (Q, i, \beta)$. We are given that φ is a homomorphism; that is, $\varphi i = \alpha$, $\varphi(q\sigma) = (\varphi q)\sigma$, $\beta q = \tilde{j}(\varphi q)$. We must show that $\tilde{\varphi}$ is a homomorphism, that is, $\tilde{\varphi} j = \beta$, $\tilde{\varphi}(\sigma r) = \sigma(\tilde{\varphi} r)$, $\alpha r = \tilde{i}(\tilde{\varphi} r)$. Computing directly we have $(\tilde{\varphi}j)(q) = (\varphi q)(j) = \tilde{j}(\varphi q) = \beta q$; $(\tilde{\varphi}\,(\sigma r))(q) = (\varphi q)(\sigma r) = ((\varphi q)\sigma)(r)$ $= (\varphi(q\sigma))(r) = (\tilde{\varphi} r)(q\sigma) = (\sigma(\tilde{\varphi} r))(q)$; $\alpha r = (\varphi i)(r) = (\tilde{\varphi} r)(i) = \tilde{i}(\tilde{\varphi} r)$. //

Definition 3.1.5*. Let $N = [R, j, \alpha]$, $A = (Q, i, \beta)$. If $\psi: N \to N(A)$ is a representation, the *dual representation* $\tilde{\psi}: A \to A(N)$ is given by the function $\tilde{\psi}: Q \to V^R$ defined by

$$(\tilde{\psi}q)(r) = (\psi r)(q). \quad //$$

Proposition 3.1.6*. The dual $\tilde{\psi}: A \to A(N)$ of a representation $\psi: N \to N(A)$ is representation. //

Proposition 3.1.7. $\varphi \mapsto \tilde{\varphi}$, $\psi \mapsto \tilde{\psi}$ are inverse bijections. //

In the language of category theory, the assignments $N \mapsto A(N)$, $A \mapsto N(A)$ can be extended to a pair of contravariant right adjoint functors between the categories of networks and automata.

Proposition 3.1.8. The representation $A \to A(N(A))$ dual to the identity $N(A) \to N(A)$ is a realization, provided that V has at least two elements.

Proof. The function in question is $q \mapsto (g \mapsto g(q))$ which may be seen to be an injection by taking g to be the characteristic function of the singleton q. //

Proposition 3.1.8*. The representation $N \to N(A(N))$ dual to the identity $A(N) \to A(N)$ is a realization. //

The construction $A \mapsto N(A)$ has been considered by Wymore (1967), and by Arbib and Zeiger (1969) who observed its categorical properties. These authors, however, interpret $N(A)$ as an automaton; that is, as a state description of a system; and so are unable to view the construction in terms of representations. This is not merely a terminological issue, since as an automaton, $N(A)$ does not have the behaviour of A. In fact, if $\rho: \Sigma^* \to \Sigma^*$ is the function sending each string to its *reversal*; i.e. $\rho 1 = 1$, $\rho(\sigma x) = (\rho x)\sigma$; then the behaviour of $N(A)$ *as an automaton* is the composition (behaviour of A) $\cdot \rho$. This has led Wymore (and others) to assert that time runs backwards in the dual system. We reject this interpretation not only because of its dubious metaphysics, and not only because we have a consistent alternative interpretation, but also on the following purely formal grounds. This entire development can be carried through in a setting in which the input monoid is replaced by an object for which there is no reversal anti-isomorphism ρ, hence there is no way to compare a left action of the input with a right action. The reader is referred to Bainbridge (1975) for this more general presentation.

Notice that although the constructions $N \mapsto A(N)$, $A \mapsto N(A)$ do not yield a satisfactory duality principle (since $A(N(A))$ is never isomorphic to A), there is a genuine duality principle for representations. This observation will be further developed in Section 3.3.

3.2. *Minimal Realizations*

Definition 3.2.1. A network $N = [R, j, \alpha]$ is *observable* if for each $r \in R$ there exists $x \in \Sigma^*$ such that $r = xj$. //

Definition 3.2.1*. An automaton $A = (Q, i, \beta)$ is *reachable* if for each $q \in Q$ there exists $x \in \Sigma^*$ such that $q = ix$. //

Definition 3.2.2. A network $N = [R, j, \alpha]$ *has distinguishable variables* if for any $r, r' \in R$, $r \neq r'$, there exists $x \in \Sigma^*$ such that $\alpha(xr) \neq \alpha(xr')$. //

Definition 3.2.2*. An automaton $A = (Q, i, \beta)$ *has distinguishable states* if for any $q, q' \in Q$, $q \neq q'$, there exists $x \in \Sigma^*$ such that $\beta(qx) \neq \beta(q'x)$. //

Property 3.2.2* is often called 'observability', but we believe that this is better reserved for the property 3.2.1, whose intuitive content is that the value of any variable may be observed as the output resulting from a suitably chosen input.

Proposition 3.2.3. If $N = [R, j, \alpha]$ is observable then $A(N)$ has distinguishable states.
Proof. Given $f, f' \in V^R$, if $f \neq f'$ let $r \in R$ be such that $f(r) \neq f'(r)$. Since N is observable, let $r = xj$. Then $\tilde{j}(fx) = (fx)(j) = f(xj) = f(r) \neq f'(r) = f'(xj) = (f'x)(j) = \tilde{j}(f'x)$. Thus $A(N)$ has distinguishable states. //

The converse of 3.2.3 is false. For finite dimensional linear systems, with $A(N)$ suitably redefined (as the Kalman dual of N), the comparable converse statement is true.

Proposition 3.2.3*. If $A = (Q, i, \beta)$ is reachable, then $N(A)$ has distinguishable variables. //

Definition 3.2.4. For any network $N = [R, j, \alpha]$, define the *reachable state description* of N to be the subautomaton $\bar{A}(N) = (Q(N), \alpha, \tilde{j})$ of $A(N)$ having states $Q(N) = \{\alpha x \mid x \in \Sigma^*\} \subset V^R$. //

Definition 3.2.4*. For any automaton $A = (Q, i, \beta)$, define the *observable coordinate description* of A to be the subnetwork $\bar{N}(A) = [R(A), \beta, \tilde{i}]$ of $N(A)$ having variables $R(A) = \{x\beta \mid x \in \Sigma^*\} \subset V^Q$. //

If we consider categories of observable networks and reachable automata, the constructions $N \mapsto \bar{A}(N)$, $A \mapsto \bar{N}(A)$ can be extended to a pair of contravariant right adjoint functors.

Definition 3.2.5. An automaton is *minimal* if it is reachable and has distinguishable states.

Definition 3.2.5*. A network is *minimal* if it is observable and has distinguishable variables.

We recall the uniqueness property of minimal automata, and assert the dual property for minimal networks.

Proposition 3.2.6. Any two minimal automata with the same behaviour are isomorphic. //

Proposition 3.2.6*. Any two minimal networks with the same behaviour are isomorphic. //

The relation between duality and minimality is the following.

Proposition 3.2.7. If N is observable then $\bar{A}(N)$ is a minimal automaton having the behaviour of N.

Proof. $\bar{A}(N)$ is by definition the reachable part of $A(N)$ and so has the same behaviour as $A(N)$. Thus by 3.1.3, $\bar{A}(N)$ has the behaviour of N.

By construction, $\bar{A}(N)$ is reachable. Since N is observable, then by 3.2.3, $A(N)$ has distinguishable states. Thus the subautomaton $\bar{A}(N)$ has distinguishable states and so is minimal. //

Corollary 3.2.8. $\bar{A}(\bar{N}(A))$ is minimal and has the behaviour of A. //

Proof. $\bar{N}(A)$ is observable and has the behaviour of $N(A)$ and hence of A.

Corollary 3.2.9. For minimal A,

$$\bar{A}(\bar{N}(A)) \cong A.$$

Proof. 3.2.8 and 3.2.6.

Corollary 3.2.10. For observable N,

$$\bar{A}(\bar{N}(\bar{A}(N))) \cong \bar{A}(N).$$

Proof. 3.2.7 and 3.2.9.

Unlike the functors $A(\ ), N(\ )$, iteration of $\bar{A}(\ )$ and $\bar{N}(\ )$ converges. In particular, for minimal automata and networks one has a genuine duality; such objects are isomorphic to their second duals.

3.3. *Coordinatized Systems*

We have argued that system theory could be based on abstract coordinate descriptions alone, just as modern system theory is based on abstract state descriptions. We propose, however, that system theory should be based not on one or the other, but on a model of a system in which neither side of the duality has a preferred status. A system should consist of a state description realized on an explicitly given abstract coordinate description which, to preserve symmetry, is realized by the given state description via the dual representation. Thus duality is built in, although in what may appear to be an artificial way. The fact that systems in the real world come with given coordinate descriptions is one line of argument that might be pursued against the charge of artificiality. There is, however, a much stronger *a priori* argument in favour of this formulation of a system: *every behaviour is realized by a canonical system of this type*.

We given for the case of automata the appropriate definitions and some of their consequences.

Definition 3.3.1. A *coordinatization of* a set Q *by* a set R with *values* in a set V is a function

$$((q, r) \mapsto qr): Q \times R \to V$$

such that the two derived functions

$$(q \mapsto (r \mapsto qr)): Q \to V^R$$
$$(r \mapsto (q \mapsto qr)): R \to V^Q$$

are injective. //

Definition 3.3.2. A *coordinatized automaton* with *inputs* Σ and *outputs* V consists of a coordinatization $Q \times R \to V$ of a set Q of *states* by a set R of *variables*, an *initial state* $i \in Q$, an *output variable* $j \in R$, a *state transition function*

$$((q, \sigma) \mapsto q\sigma): Q \times \Sigma \to Q$$

and an *information flow function*

$$((\sigma, r) \mapsto \sigma r): \Sigma \times R \to R$$

which satisfy

$$(q\sigma)r = q(\sigma r). \quad //$$

Definition 3.3.3. Referring to 3.3.2, define $A = (Q, i, \beta)$ where $\beta: Q \to V$ is given by $\beta(q) = qj$. Define $N = [R, j, \alpha]$ where $\alpha: R \to V$ is given by $\alpha(r) = ir$. Define $\varphi: Q \to V^R$ by $(\varphi q)(r) = qr$. Then $\varphi: A \to A(N)$ is a homomorphism, since $(\varphi i)(r) = ir = \alpha(r)$; $\varphi(q\sigma)(r) = (q\sigma)r = q(\sigma r) = (\varphi q)(\sigma r) = ((\varphi q)\sigma)(r)$; $\beta(q) = qj = (\varphi q)(j) = \tilde{j}(\varphi q)$. Moreover, φ is injective, by the definition of a coordinatization. Likewise $\tilde{\varphi}: N \to N(A)$ is injective. Thus a coordinatized automaton is equivalent to an automaton, a network, and a pair of dual realizations of each on the other. The automaton A is called the *state description* of the coordinatized automaton, and N is its *coordinate description*. The coordinatized automaton is denoted $A \to A(N)$; or dually, $N \to N(A)$. //

Example 3.3.4. The following data specify a coordinatized automaton

	v	w	x	y	z
a	0	0	1	1	1
b	0	1	0	1	0
c	0	0	1	0	0
d	0	0	1	0	1
e	0	0	0	0	0

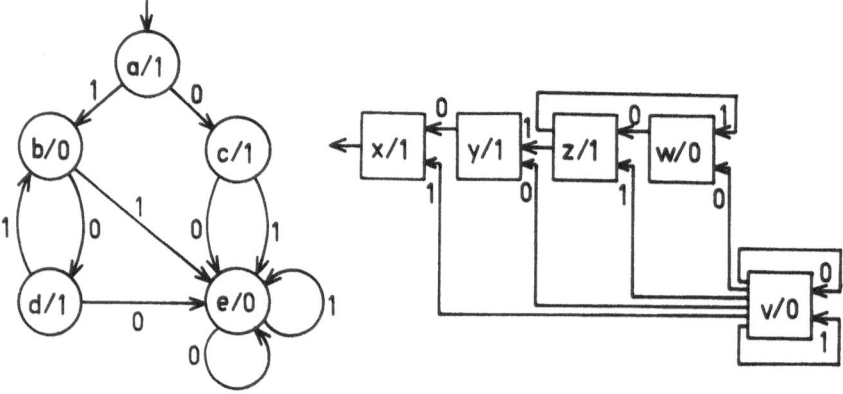

The diagram on the left is a *state transition diagram*. The circles are states and bear labels (name)/(output) which tabulate the output function. The initial state is designated by an incoming arrow with no source. The state transition function is represented by the arrows and their labels. The diagram on the right is the information flow diagram of Example 2.1. The table is the coordinatization of states by variables. Note that the output function also appears as the x-column of the coordinatization, and the values function of the network appears as the a-row of the table. //

By 3.1.8, any automaton can appear as the state description of a coordinatized automaton. By 3.1.8*, every network can appear as the coordinate description of a coordinatized automaton. The facts which most strongly justify the introduction of the notion of coordinatized automaton are the following.

Theorem. 3.3.6 For any minimal automaton A, there is a canonical coordinatized automaton whose state description is A and whose coordinate description is $\bar{N}(A)$.

Proof. The dual of the realization $\bar{N}(A) \to N(A)$ is a homomorphism $A \to A(\bar{N}(A))$ whose image is $\bar{A}(\bar{N}(A))$. Since A is minimal, $A \cong \bar{A}(\bar{N}(A))$, so $A \to A(\bar{N}(A))$ is a realization. This gives the required coordinatized automaton. //

Theorem 3.3.7. For every function $f: \Sigma^* \to V$ there is a canonical coordinatized automaton $A \to A(N)$ is which both A and N are minimal and have behaviour f.

Proof. Given $f: \Sigma^* \to V$, we may construct a minimal automaton with behaviour f as follows. Define $A_f = (\Sigma^*, 1, f)$, with multiplication in Σ^* as transition function. The behaviour of A_f is $x \mapsto f(1x) = f(x)$. By 3.2.8, $\bar{A}(\bar{N}(A_f))$ is minimal and has behaviour f. By 3.3.6 we obtain the desired coordinatized automaton, noting that $\bar{N}(\bar{A}(\bar{N}(A_f)))$ has behaviour f. //

We can now give the most general statement of the result on series parallel decomposition suggested by Example 2.4. Theorem 3.3.10 generalizes some results of Wymore (1967) on parallel decomposition.

Definition 3.3.8. A *transition congruence* on an automaton $A = (Q, i, \beta)$ is an equivalence relation π on Q such that for all $\sigma \in \Sigma$, $q, q' \in Q$,

$$q \pi q' \text{ implies } (q\sigma)\pi(q'\sigma). //$$

Definition 3.3.9. A set S of variables of a network $N = [R, j, \alpha]$ is *stable* if for all $\sigma \in \Sigma$

$$s \in S \text{ implies } \sigma s \in S.$$

Theorem 3.3.10. For any coordinatized automaton $A \to A(N)$ there is a Galois connection between transition congruences on A and stable sets of variables in N.

Proof. For equivalence relations π, ρ on a set Q, write $\pi \leqslant \rho$ if for all $q, q' \in Q$

$$q \pi q' \text{ implies } q\rho q'.$$

Relative to a coordination $Q \times R \to V$, for $r \in R$ write $qr^{\#}q'$ to mean $qr = q'r$.

For an equivalence relation π on Q, define the subset $\pi^{\#}$ of R by

$$\pi^{\#} = \{r \in R \mid \pi \leqslant r^{\#}\}.$$

For a subset S of R, define the equivalence relation $S^{\#}$ on Q by

$$qS^{\#}q' \text{ if and only if } qs^{\#}q' \text{ for all } s \in S.$$

This defines a Galois connection between subsets of R and equivalence relations on Q; that is

$$\pi \leqslant \rho \quad \text{implies} \quad \rho^{\#} \subset \pi^{\#}$$
$$S \subset T \quad \text{implies} \quad T^{\#} \leqslant S^{\#}$$
$$\pi \leqslant S^{\#} \quad \text{if and only if} \quad S \subset \pi^{\#}.$$

Each of these is an immediate consequence of the definitions.

Suppose that π is a transition congruence on A. Then $\pi^{\#}$ is stable, as follows. Let $r \in \pi^{\#}$ and $\sigma \in \Sigma$. If $q \pi q'$ then $(q\sigma)\pi(q'\sigma)$, so $(q\sigma)r = (q'\sigma)r$. Thus $q(\sigma r) = q'(\sigma r)$, i.e., $q(\sigma r)^{\#} q'$, so $\pi \leqslant (\sigma r)^{\#}$. Thus $\sigma r \in \pi^{\#}$ and so $\pi^{\#}$ is stable.

Suppose that S is a stable set of variables of N. Then $S^{\#}$ is a transition congruence, as follows. Let $q S^{\#} q'$ and $\sigma \in \Sigma$. If $s \in S$, then $\sigma s \in S$, so $q(\sigma s)^{\#} q'$; that is $q(\sigma s) = q'(\sigma s)$. Thus $(q\sigma)s = (q'\sigma)s$; that is $(q\sigma)s^{\#}(q'\sigma)$. Thus $(q\sigma)S^{\#}(q'\sigma)$, so $S^{\#}$ is a transition congruence. $//$

Remark 3.3.11. One might well want to generalize the definition of coordinatized automaton to permit several output variables and several initial states. We have chosen not to do this for simplicity of exposition. Details of this and other generalizations may be found in Bainbridge (1972), (1975).

4. CONCLUSION

A proper understanding of this duality requires that two facts be stated clearly. The first is that the duality applies in its most general form not to systems but to *representations* of systems in the sense of Definitions 3.1.4 and 3.1.4*. The second is that the duality should be interpreted as a translation from the language of state transition to the language of information flow; the result of the construction $A \rightarrow N(A)$ is a coordinate description of the *same* system, not a state description of some other system (the 'dual' system). Failure to recognize one or both of these points has, in our opinion, led to some confusion in the literature and to the failure of the fundamental significance of this duality to be appreciated.

ACKNOWLEDGEMENT

This work was supported by grant NSF DCR 72-03703A01 at Columbia University.

REFERENCES

Arbib, M. A. and Manes, E. G.: 'Adjoint Machines, State Behaviour Machines, and Duality', *J. Pure Appl. Alg.* to appear.
Arbib, M. A. and Zeiger, H. P.: 1969, 'On the Relevance of Abstract Algebra to Control Theory', *Automatica* 5, 589–606.
Bainbridge, E. S.: 1972, 'A Unified Minimal Realization Theory with Duality', Ph.D. Dissertation, Computer and Communication Sciences, University of Michigan.
Bainbridge, E. S.: 1975, 'Addressed Machines and Duality', *Lecture Notes in Computer Science*, No. 25, pp. 93–98, Springer-Verlag, New York.
Hartmanis, J. and Stearns, R. E.: 1966, *Algebraic Structure of Sequential Machines*, Prentice Hall, N.J.
Kalman, R. E.: 1960, 'On the General Theory of Control Systems', *Proc. First IFAC Congress, Moscow*, Butterworth, London.
Wymore, A. W.: 1967, *A Mathematical Theory of Systems Engineering: The Elements*, Wiley, New York.

CHAPTER 4

TOWARDS A SYSTEMS METHODOLOGY OF SOCIAL CONTROL PROCESSES

WALTER BUCKLEY

In recent years it has become permissible, if not mandatory, for systems theorists to apply analogies across widely different fields. I intend to engage in this to some extent today, in a very tentative and exploratory way, with the hope of provoking constructive responses from the splendid array of participants at this Symposium. I will also outline the results of some work with an engineering colleague which attempts to develop a systems methodology for studying social control processes, with the same hope of 'positive' feedback.

I begin, then, with the question, is it possible or feasible to apply the recent work in non-equilibrial or irreversible thermodynamics to the study of social structures and processes? There is a long history of the use of the equilibrium concept in sociology, but that has always seemed to me and many others to be unhelpful and misleading, focusing as it did on the stability problems of an apparently closed or isolated system. The newer work focuses on open, dissipative systems and the creation of structure, and provides an opening for some of the insights of information and communication theory and their relation to entropic processes.

Glansdorff and Prigogine, in their book – *Thermodynamic Theory of Structure, Stability and Fluctuations*, Wiley, New York (1971) state:

> It is a rather remarkable coincidence that the idea of evolution emerged in the nineteenth century associated with two conflicting aspects: In thermodynamics, the second law is formulated as the Carnot–Clausius principle. It appears essentially as the evolution law of continuous disorganization, i.e., of disappearance of structure, introduced by initial conditions.
>
> In biology or in sociology, the idea of evolution is, on the contrary, closely associated with an increase of organization giving rise to the creation of more and more complex structures.

The authors reconcile the apparent contradiction, of course, by arguing that though there is only one type of physical law, there are different thermodynamic situations – those near equilibrium, and those far from it.

> Broadly speaking *destruction of structures* is the situation which occurs in the neighbourhood of thermodynamic equilibrium. On the contrary, *creation of structures* may

W. E. Hartnett (ed.), Systems: Approaches, Theories, Applications, 63–70.

occur, with specific non-linear kinetic laws beyond the stability limit of the thermodynamic branch. This remark justifies Spencer's point of view (1862): "Evolution is integration of matter and concomitant dissipation of motion".

The application of such a thermodynamics of dissipative structures to eco-systems seems the most suggestive analogy for sociocultural systems. Firstly, there is a dependence on flows of energy, matter, and information. Secondly, there is the process of 'succession', whereby characteristics develop to some stable state and are only weakly determined by external conditions but strongly by the properties of the structures themselves. Then there may be attainment of 'climax states' where a maximum of matter (biomass or socio-mass) has been entrained into cyclic flows. Finally, such systems show homeostatic properties, with a more and more delicate balance to be maintained among the components. As Thomas Blackburn points out in his article in *Science* (21 Sept., 1973), the increasing selective advantage of behavioral control of mass and energy flows accounts for the successional trend toward greater structure and higher efficiency, but at the cost of net productivity because of the energy cost of maintaining that structure.

When we come to translate this sort of thing into sociocultural analysis, there are obvious difficulties. In addition to the need to introduce a sub-stantial role for information processes, we need to find an equivalent for energy flow. The most likely candidate is some kind of motivational or psychological energy as the driving force of strictly social relationships. D. J. McFarland, for one, has pointed to the wide but loose use of a motivational energy concept in the behavioral sciences, and derives a more precise meaning from a generalized effort variable which he applies in a control model of animal behavior in the regulation of physiological variables. [In Daniel S. Lehrman, *et al*. New York, (eds.), *Advances in the Study of Behavior*, Vol. 3, Academic Press, New York, 1970]. Unfortunately, we are not ready for this degree of mathematical precision in social science.

My present very rough conception of a social non-equilibrial thermodynam-ics is based on the following considerations. The human *social* level develops out of the mental coupling, as it were, of individuals and their motivational states, which leads to the patterned regularities of interaction constituting the phenomenon of social organization or structure. This patterning is based on common understandings or more or less generalized rules – the norms and values – which become instituted in social roles and positions. These are dissipative structures requiring continuous inputs of energy, especially motiva-tional energy, matter, and information. Once developed, social structures are supported by the commonly accepted rules, which generate the more detailed behaviors and interactions representing the flow of social processes occurring within the structural framework. Maintenance of the structure requires that the role behaviors — which require continuous motivational energy — lead to outcome that 'feed', i.e., reinforce this energy. Social structures are, of course, goal-seeking, decision-making structures of a hierarchical nature in which feedbacks promote goal-attainment on each hierarchical level (sometimes in a very unbalanced way), as well as the homeostatic properties that tend to perpetuate the structures and the component elements.

Once a structure is developed, it tends to 'entrain' individuals and groups within its constraints and its benefits. Once developed, if it is of any size and complexity, such a structure takes on a life of its own and resists change. A move to a new structure with a higher adaptive capability relative to its external and internal environment requires a large input of new motivational energy, and the new information, ideas, and communication nets required to recode the rules which are to define the newly created structures.

But social structures do have their inherent morphogenic properties inducing structural change, and the main eventual question is whether it occurs destructively or constructively.

This brings me to the discussion of recent work with an engineer friend and colleague, Dr. Misha Pergler, who has brought his control engineering skills to bear in helping to develop a systems methodology for modelling social control or regulation processes on the broader societal level.

We start with the assumption that social systems have evolved over time, and — like biological systems — their self-regulating structures and processes have evolved as well, with varying degrees of success in maintaining the viability of the system. Although small, primitive societies might be seen as quite effective self-regulating systems, it can be argued that large-scale societies, especially our industrialized—urbanized sociocultural systems, are only just beginning to evolve the self-regulating structures demanded for longer-run viability in the face of the internal system complexities and external environmental challenges. From a cybernetics point of view, modern societies are rather poor self-regulators, especially in terms of longer run goals and global viability. The 'muddling' approach of many complex societies has led to outcomes neither intended nor desired by most member individuals or groups: urban blight, squandered resources, plundered environment, runaway technology, moral decay of political and economic institutions, paranoid bellicose orientations toward other nations, etc.

(By the way, I use the term 'self-regulation' here in a loose and intuitive way, recognizing the arguments that no system is such in any precise sense.)

Given the problems of regulation in sociocultural systems, we can argue, however, that the historical shift from monarchical or oligarchical types of control or regulating system to the more or less democratic types can be seen as an important step in the evolution of a better self-regulating social system (in a sense to be defined), though it introduces new system problems of its own. (For example, in a democratic system there is potentially fuller information flow, both about internal states of the system and about external disturbances. Also there are potentially more levels or metalevels of regulation, down to the level of the individual and subgroup, with greater autonomy at lower hierarchical levels and fuller feedback between levels in each direction. As Herbert Simon and others have argued, the more complex the system, the greater the number of hierarchical levels of subsystem it tends to develop, as if complexity at any one level can only become so great if viability is to be promoted. Also, decision-making in more democratic systems has a greater breadth of knowledge and strategy alternatives to draw on. And so forth.)

It is next argued that two of the essential regulative processes in living systems are morphostasis or self-stabilization (MS) and morphogenesis or self-organization (MG). If homeostasis is defined as the regulation of critical system variables to within certain limits, morphostasis — as implied literally — may refer to regulation of the structure — expecially the control structure — of the system to within certain limits. In other words, MS refers to processes tending to prevent significant changes in the organization of the system, or the *status quo*, if you wish. Any system must maintain the integrity of its structure with some minimal time invariance, depending on the challenges of the environment.

But given severe enough challenges of the external or internal environment, the viability of a system depends on its ability to change its operating structure — to adapt so as to improve goal attainment or regulate against 'disturbances'. This is how we understand the process of species evolution. We refer to this as morphogenesis, again using the word in its more literal sense.

But now we must make a distinction between at least two types of MG. Throughout the history and prehistory of sociocultural systems we take it that most of the significant structural changes, i.e., MG, have taken place along with a high level of destructive conflict: civil war, revolution, conquest. We will call this, "morphogenesis via destructive conflict", or MG_1. This is a high risk type of structural change, increasingly so as social systems became more complex and more powerful.

In recent historical times a different kind of institutionalized regulatory structure has begun to emerge promoting structural changes based on a competition of ideas and joint decision-making. We may call this MG_2. This is the more democratic type of system, in which institutional principles incorporate *explicit procedures* for self-modification (e.g., amendments) and hence for change of the structure of the system based on such principles. Such a system, however, still has a long say to go to evolve into its full MG_2 potential.

We may hypothesize that as larger-scale, more complex socioculture systems evolved, morphostatic mechanism became more thoroughly institutionalized in them compared to MG_2 mechanisms. (This could be argued in terms of the (non-equilibrial) thermodynamics of sociocultural systems: the greater input of information of negentropy to generate the meta-levels of interpersonal coupling and societal integration required for MG_2 regulation.) Modern societies are thus seriously unbalanced in terms of MS_1 versus MG_2 processes. The result is to drive the system toward more and more crises and the generation of destructive MG_1 processes.

Part of the research goal is to investigate the additional institutionalized structures and processes needed to generate morphogenesis MG_2, and soften the morphostasis. This involves, among other things, (1) more explicit and effective procedures for defining and operationalizing collective goals; (2) procedures for adequately defining the state of the system at any time and its probable trajectories; (3) organizations for generating scientifically informed alternative policy strategies, effective decision-making procedures, and implementation of the chosen strategy; (4) effective citizen participation; (5) better mechanisms for feedback of goal-deviating information. All of this implies the development of much better models of the structure and operation of complex social systems; contemporary social science has hardly made a dent in this, and research support at least the size of that for high-energy physics is demanded.

The foregoing is a very informal sketch of the main thrust of our work. Since we have attempted to develop a more precise systems methodology based on the engineering approach of Dr. Pergler, some of you may be interested. For pedagogical and epistemological reasons, the starting point is the development of a basic set of operationalized concepts, the more complex ones defined in terms of the simpler ones. A more or less standard input—output block symbolism is used. The basic unit is the 'element', defined as a model of an object of the external world (or occasionally as an abstract object or relation). The element has one or more input and output variables, and a 'characteristic', or mathematical function, defining the relationship between the two — whether continuous or discrete, deterministic or probabilistic. The initial concept is that of 'association', defined as an element and associated variables such that the knowledge of one variable decreases the uncertainty of prediction of the other (Figure 1). 'Causation' or causal

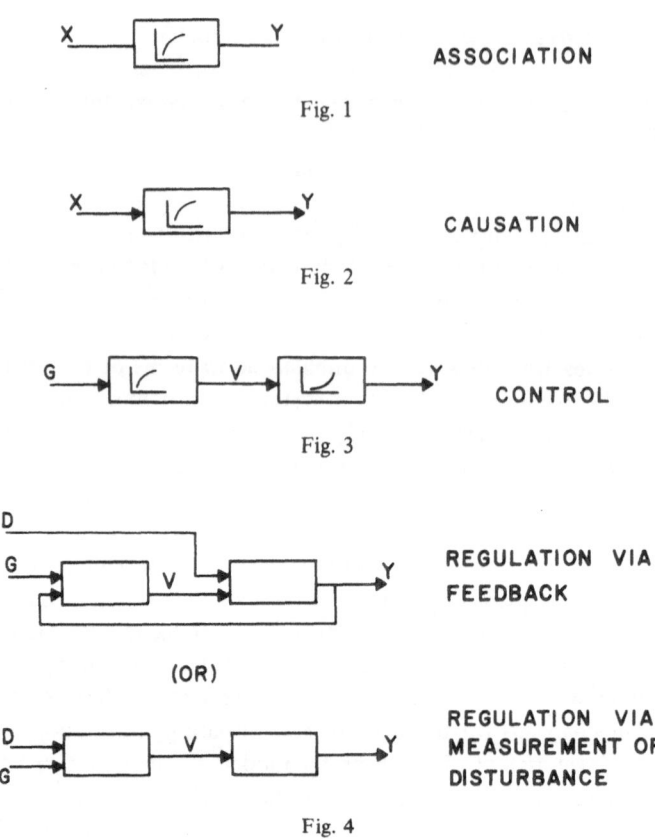

Fig. 1   ASSOCIATION

Fig. 2   CAUSATION

Fig. 3   CONTROL

REGULATION VIA FEEDBACK

(OR)

REGULATION VIA MEASUREMENT OF DISTURBANCE

Fig. 4

relation is an association of variables whereby a change of the input variable is followed by a change of the output, the change being described by the characteristic of the element (Figure 2). 'Control' is defined as two causal elements; the input to one, the 'controlling-element', is a 'goal' input. Its output is a connecting variable which is identical to the input of the second element, the controlled element, and constitutes a 'structural linkage' between them. If the output variable shows an association with or matching of the goal input when the structural linkage is intact, but fails to do so when the latter is broken we speak of a control system, or sometimes, a 'control channel'. (Figure 3).

A 'regulation' system is a control system in which the goal can be achieved by different trajectories of the connecting variables, in response to 'disturbance' inputs to the system. In feedback regulation, disturbances are inputs into the controlled element and some of the output is fed back into the control element. In "regulation via measurement of the disturbance", disturbances are inputs into the control element directly. In both cases the quality of regulation is measured by the degree of association between the output and the goal input (Figure 4).

A 'system' is thus a model consisting of a universe of elements with their characteristics and inputs and outputs, and a structure consisting of the structural linkages as defined above.

'Adaptation' of a system becomes a change in the overall characteristic of a regulator that increases the quality of regulation of the system compared to that without the change. If the regulator is decomposed into its system elements and structure, adaptation may be found to involve a change in (a) the characteristics of one or more of the elements, and/or (b) the structure of the system, and/or (c) the universe of elements of the system.

'Homeostasis' is defined as the regulation of critical conditions necessary to the continued functioning of the system. 'Morphostasis' refers to homeostatic regulation in which the very structure of the system itself is taken as the goal of regulation. That is, the structure tends to be maintained unchanged despite forces or 'disturbances' tending to change it.

Finally, 'morphogenesis' is defined as adaptive change of the structure of the system; that is, a change of structure that results in better goal attainment or regulation of the system.

In applying such a framework to social control or regulatory processes, the basic elements of the model are decomposed into generalized decision-making elements. Thus, for example, the decision output of a controlling element becomes an input to the controlled element. Our decision-making element, or decision channel, is intended as a model of a person or more usually a social organization of persons. It incorporates the several stages in the decision process accepted more or less in modern decision theory. The input to the decision channel is sensory data from the external world, especially linguistic messages. The first element in the channel selects certain of these data to attend to and outputs these to an element that selects a problem requiring decision. (All of these elements are interconnected with an accumulated 'knowledge' element and an element representing the values or 'goal space' of the modeled individual or group). The third element selects a set of relevant strategies and a fourth predicts the probable outcomes of each. The fifth evaluates the outcomes in terms of the decision-maker's goal-space, the sixth selects the resulting best strategy and the last selects the initial action for implementing that strategy. The output of the decision-channel is thus that initial action.

If we interconnect two or more such components as elements in a social regulatory system, we may attempt to develop models of how we conceive the regulatory processes of complex societies to operate in actuality, and models of how we might improve such regulatory systems to promote, for example, MG_2 as needed. Thus, we might develop alternative models of contemporary American society according to a 'conservative' version, a 'liberal version', or a 'radical version', depending on how those theories visualize the structure of the regulatory system, the universe of elements, and their characteristics. In this way we may promote a constructive scientific critique of various models of actual society and of models of how it might be improved as a viable system. It is my opinion that, especially in the context of the current ecosystem movement, there is already a good deal of activity moving in this direction. The work outlined here might be seen as a modest exploratory step in the direction of the development of a system methodology for promoting this movement. Science itself is a morphogenic system and thrives on constructive critique.

CHAPTER 5

STATES AND EVENTS

MARIO BUNGE

1. INTRODUCTION

Whereas the concepts of state, event, and process belong to ordinary knowledge, those of state variable and state space seem to have sprung up in thermodynamics and statistical mechanics. From these fields they spilled over
into others, such as electrical engineering, psychology, biology, automata
theory, and systems theory. Systems theorists have of course paid close attention to the technical concepts of state variable and state space, endeavour-
ing to elucidate them in the most general and accurate way possible.
However, there is room to doubt whether the usual accounts of these concepts, given in the systems theory literature, are indeed as general as claimed
– e.g., whether they make room for continuous systems such as fields, or for
quantum-mechanical systems.

Moreover something important is missing in the accounts of state variables
and state spaces found in the general systems literature, namely the notion
of a reference frame. Although this concept has no place in automata theory,
electrical network theory, and other theories, it does occur in every explanation of the working of an electric motor – not to speak of the underlying
sciences of mechanics and electrodynamics. The notion of a reference frame
is in fact central to physics since the states of any real system are always
relative to some frame or other: just think of the state of motion of a body,
or of the thermodynamic state of a fluid, or of the state of an electric field.

A second concept one misses in general systems theory is that of a law. To
be sure a general theory of systems should neither presuppose nor contain
any specific (e.g., chemical or biological) law, for otherwise it would not
quite be general – or, equivalently, the systems it describes could not have a
number of alternative realizations employing materials of different kinds
(hence satisfying different laws). However, the general concept of a law – as
opposed to a particular law statement such as Ohm's – ought to play some
role in general systems theory if only for a mathematical reason. Indeed, but

W. E. Hartnett (ed.), Systems: Approaches, Theories, Applications, 71–95.
Copyright © 1977 by D. Reidel Publishing Company, Dordrecht-Holland.
All Rights Reserved.

for the laws — which place restrictions upon the ranges of the state variables — we should accept the usual characterization of state spaces as vector spaces, or even inner product spaces. The existence of laws rules out this characterization. In fact, if a law restricts the range of a variable, as it usually does, then it is no longer true that the product of a variable by an arbitrary scalar belongs to the same space.

Finally, the notions of a state space and a trajectory in such a space seem not to have been fully exploited in the general systems literature. For one thing, they might be employed for defining the very concept of a concrete system as opposed to a mere aggregate or conglomerate — namely as a thing whose state space is not just the union or sum of the state spaces of its components. For another, states ought to serve to define events, namely as changes of state, or, equivalently, the concept of a state space ought to occur in the definiens of the concept of an event space — a concept which seems to have gone unnoticed in general systems theory. Thirdly the notion of a state space seems to be the most natural matrix where the difference between a history (or life line or line of behaviour) and a law could be elucidated.

This paper is addressed to the above-mentioned problems. It may therefore be placed in the foundations of the general theories of systems. And, since it is foundational, it overlaps with philosophy. Indeed, the branch of philosophy known as ontology (or metaphysics, or philosophical cosmology) deals with things of all kinds, their states, and their changes of state. However, most philosophers — like most engineers and scientists — do not bother to analyze these basic concepts — not even when, as in the case of process metaphysics, these occur centrally in their own systems. (Typically, Hegel, Bergson, Whitehead and their numerous followers have failed to clarify the notion of an event, basic to their metaphysics. Hence they have bequeathed to us utterly obscure philosophies of change.)

The concepts of state and event are employed not only in ontology but also in epistemology — the theory of knowledge —, but uncritically since they are not analyzed. For example the concept of a state occurs in the notion of a state description, or a proposition purporting to describe a state of affairs. And yet the very concept of a state goes unanalyzed in exact and influential — even though non-scientific — philosophies, such as Carnap's and the possible worlds metaphysics, that make ample use of it.

Such a lack of analyses is deplorable if only because the concept of a state, and hence the notions of a state description and a change of state, pose interesting philosophical problems. For example, since a state is determined by state variables, and since the choice of the latter is partly determined by the state of the art, what right do we have in assigning a thing an objective state portrayed by a state description? Another example: the so-called identity theory (or rather hypothesis) is usually formulated as a statement of the (nonlogical) identity of mind and brain. Would the hypothesis not gain in precision by construing it as a statement that mental states are states of the nervous system (or rather of subsystems thereof), and that therefore mental events are changes in the states of neuron assemblies?

So much for the motivations of our study. And now, before turning to it, an important caveat. The concept of state variable (and hence that of state space) introduced in this paper is in the tradition more of physics and biology than of systems theory where a far more restricted concept is employed. In fact, in much of the latter the state variables are those mediating between inputs and outputs: they are what psychologists term 'intervening variables'. No such restriction will be placed on the concept used in this paper, for the very distinction between inputs and outputs, so natural in technology (particularly in control theory), is rather artificial with reference to natural systems such as an atom or a family.

2. PROPERTIES AND PREDICATES

We shall be dealing with concrete things of any kind – physical, chemical, biological, social, or technical. When a thing is complex and its components are coupled it is called a *system*. Each system component possesses certain properties, and the system as a whole has further properties that its components may not possess – for example, composition, stability, and mode of disintegration. Any such properties, when known or suspected, are conceptualized as attributes (predicates, statement valued functions).

Needless to say no such conceptualization need be unique: one and the same property may be represented by alternative concepts. Thus for different physical theories alternative representations of the energy of a thing may be adopted, and for different sociological theories alternative measures of participation may be chosen. Moreover not all predicates represent properties of things. In particular negative predicates, such as 'featherless', and disjunctive predicates, such as 'interacting or conducting', represent no properties of things even though they may occur in our discourse about things. Let these crude and dogmatic remarks suffice for introducing the notion of a thing property as distinct from a predicate.

Each property P of a thing can be conceptualized or represented by some function F from a certain domain A to some codomain B. We shall abbreviate:

F represents P as $F \doteq P$, where F maps A into B, i.e., $F: A \rightarrow B$.

The nature of A, B and the correspondence F between them will of course depend upon the property P being represented. A few rather typical examples will suggest the rich variety of property-representing functions we shall have to cope with, and will prepare the ground for a general definition.

Example 1. Dichotomic global property. Let $F \doteq$ Stability. Then,

> A = Set of all concrete systems (physical, chemical, biological, or social)
>
> B = Set of all propositions of the form "x is stable", where $x \in A$.

Example 2. Qualitative global property. Let $F \doteq$ Social structure. Then $A = S \times T$, where:

> S = Set of all human societies
>
> T = Set of instants of time
>
> B = Family of all sets of persons.

and, for $s \in S$, $t \in T$, $F(s, t)$ = Family of all social groups included in society s at time t.

Example 3. Quantitative global property. Let $F \doteq$ Electric charge. Then $A = \mathscr{B} \times T \times U_{\chi}$, where:

> $\mathscr{B}$ = Set of all bodies
>
> T = Set of instants
>
> U_{χ} = Set of all electric charge units
>
> B = $\mathbb{R}$ = Set of all real numbers

so that '$F(b, t, u) = r$', for $b \in \mathscr{B}$, $t \in T$, $u \in U_{\chi}$, and $r \in \mathbb{R}$, abbreviates 'The electric charge of body b, at time t, expressed in unit u, equals r'.

Example 4. Quantitative global stochastic property. Let $F \doteq$ Momentum probability distribution. Here $A = Q \times \mathscr{F} \times T \times \mathbb{R}^3$, where:

> Q = Set of quantum-mechanical entities of a certain kind
>
> $\mathscr{F}$ = Set of all reference frames
>
> T = Set of all time instants
>
> B = $\mathbb{R}$

so that '$F(q, f, t, p)dp$' abbreviates 'The probability that entity q, relative to frame f, at time t, has a momentum comprised between p and $p + dp$'. (In standard notion, $F = |\varphi|^2$, where φ is the Fourier transform of the state function ψ.)

Example 5. Quantitative local property. Let $F \doteq$ Gravitational potential. Then $A = \mathscr{G} \times \mathscr{F} \times E^3 \times T \times U_\epsilon$, where:

$$\mathscr{G} = \text{Set of all gravitational fields}$$
$$\mathscr{F} = \text{Set of all reference frames}$$
$$E^3 = \text{Euclidean three space}$$
$$T = \text{Set of all time instants}$$
$$U_\epsilon = \text{Set of all energy units}$$
$$B = \mathsf{R} = \text{Set of all real numbers}$$

so that '$F(g, f, x, t, u) = r$' abbreviates 'The (scalar) gravitational potential of field $g \in \mathscr{G}$, relative to frame $f \in \mathscr{F}$, at point in space $x \in E^3$, at time $t \in T$, expressed in energy unit $u \in U_\epsilon$, equals r'.

In each case the function F represents a property of entities of some kind, or a *general property* (such as age). And the value of F at a particular entity or thing is an *individual property*, or a property possessed by the individual in question (such as being 1 year old). Some properties, or rather property representing functions, are of special interest to us: they are those determining the states a thing can be in. For example, the mass density, stress density and force density in a mechanical system determine its stability properties. They are therefore called *state variables* or, better, *state functions*. On the other hand stability is not a state function but a sort of outcome of the interplay of certain state functions. Nor are t (time), x (position), f (reference frame) and u (unit of some kind) state functions – nor, indeed, functions of any kind. They are arbitrary members of certain sets and are not possessed by any thing in particular: on the contrary they are rather public in the sense that they can be 'used' by a number of things.

In Example 2 above the function F_s: $\{s\} \times T \rightarrow B$ is a state function for the individual system s. In Example 3, the function F_{bu}: $\{b\} \times T \times \{u\} \rightarrow \mathsf{R}$ is a state function for b. In Example 4, the function $F_{qf} = \{q\} \times \{f\} \times T \times \mathsf{R}^3 \rightarrow \mathsf{R}$ is a state function for q. And in Example 5, the function F_{gfu}: $\{g\} \times \{f\} \times E^3 \times T \times \{u\} \rightarrow \mathsf{R}$ is a state function for g.

The preceding remarks suggest the following preliminary characterization: A function is a *state function* for a thing of a given kind only if it represents a property possessed by the thing. Whether this representation is faithful (true) is immaterial to the function's qualifying as a state function. What is decisive is that the function refer to the thing and be interpretable as representing or conceptualizing the intended property.

3. DEFINITION OF A STATE FUNCTION

Every scientific theory concerns some species of concrete things, and every general theory of systems is about some wide genus of concrete and complex things. Every such theory, whether specific or generic, employs a finite number of state functions to describe its referents. Since some scientific theories – such as Maxwell's electromagnetism, Einstein's gravitation theory, and Dirac's electron theory – are true to a remarkable degree, it is reasonable to assume that the things they concern do in fact possess only a finite number of general properties. (Recall that we distinguish a general property, such as being electrically charged, from a particular property, such as having so many units of electric charge – just as we distinguish a function from each of its values.)

Example. Population is a conspicuous biological and sociological property. It may be conceptualized as a state function $P: S \times T \to \mathbb{N}$ from pairs ⟨community s of organisms of a kind, instant t⟩ to the natural numbers. Each value $P(s, t) = n$, for $n \in \mathbb{N}$, represents an individual property of s, whereas the function P itself represents a general property, or a property of all the members of the set S. (Shorter: P is universal in S.)

Let us now study the whole bunch of state functions for things of a given kind. In principle they have little in common except their uniform reference to things of a certain kind. In particular they need not even be defined on the same domain. However, a harmless trick will assign them all the same domain. Thus if $F_1 : A \to B$ and $F_2 : C \to D$, where $A \neq C$ and $B \neq D$, we can adopt the new state functions.

$$F_1^*: \ A \times C \to B \ \text{ such that } \ F_1^*(a, c) = F_1(a)$$
$$F_2^*: \ A \times C \to D \ \text{ such that } \ F_2^*(a, c) = F_2(c)$$

for all $a \in A, \ c \in C$.

But we usually need not resort to this trick because as a matter of fact most state functions for a thing are defined on the same domain. Think of the collection of property-representing functions (such as mass density, velocity and field potential) concerning a continuous medium such as a fluid or a field: they may all be construed as sharing a single domain, namely some four-manifold. Likewise the set of dynamical variables ('observables') of a quantum mechanical system are operators on the Hilbert space of the system. But, whether or not this is in fact the case in a given instance, the above-mentioned procedure will do the trick of uniformizing the domain of the state functions for a given thing. Therefore we can make:

Definition 1. Let $F_i: A \rightarrow V_i$, with $1 \leqslant i \leqslant n$, be a set of state functions for a thing. Then the function $F = \langle F_1, F_2, ..., F_n \rangle: A \rightarrow V_1 \times V_2 \times ... \times V_n$, such that $\langle F_1, F_2, ..., F_n \rangle(a) = \langle F_1(a), F_2(a), ..., F_n(a) \rangle$ for $a \in A$, is called a (total) *state function* for the thing. And if all the V_i are vector spaces, then F is called a *state vector*.

Notice the cautious indefinite article instead of the definite. The reason is that there is no such thing as *the* state function for a given thing: indeed there are as many state functions as representations (or models) of the thing can be conceived, i.e., any number of them. (For example, whereas Lagrangian theories employ generalized coordinates and velocities as the basic state functions, Hamiltonian theories use generalized coordinates and momenta.) And even one and the same representation is compatible with infinitely many choices of reference frames, every one of which will ensue in a different state function.

The sole test for the suitability of a choice of state functions is the adequacy (truth) of the model or theory as a whole — in particular that of its key formulas, which are those interrelating the various state functions, namely the law statements of the theory. Still there may be alternative though basically equivalent formulations of one and the same theory. (For example, most field theories can be formulated using either field strengths or field potentials, and the latter are mutually equivalent modulo certain arbitrary constants or functions.) In the case of equivalent theories there are no preference criteria other then those of computational convenience, or else heuristic power, or even sheer beauty or fashion.

To put it negatively: The choice of state functions is not uniquely determined by experimental data but depends partly upon our total knowledge, as well as upon our abilities and goals, and even our inclinations. This consideration will play an important role in any talk of states and state spaces, on which there is more in Section 6. But, lest we have given the impression that the choice of state functions is totally arbitrary and a matter of sheer taste, let us hasten to state that, whatever the set of state functions chosen, they are supposed to abide by the law statements included in the theory, this being a matter not of convention. Neither is the choice of law statements arbitrary: such statements are supposed to be reasonably true to fact. More on this in the next Section.

4. LAW STATEMENTS

The laws of reality, be they natural or social, are supposed to be objective patterns or constancies with large scopes, namely entire species of things or even genera of such. Every such pattern can be conceptualized in a number of ways — in fact in as many ways as there are choices of state functions. (For example, Maxwell's equations can be written in terms of field strengths or of potentials of various kinds.) Such conceptualizations are called *law statements*.

A law statement may be regarded as a condition on certain state functions for some thing — not an arbitrary condition of course but one tallying with similar conditions (law statements) and moreover one that has been verified (or rather confirmed) to a reasonable degree with the help of observations, measurements, or experiments. Such conditions may take a number of forms, depending not only on the thing itself but also on the state of knowledge. All of the following are conspicuous simple forms:

$R_F = V$, where F is a state function, R_F its range, and V some well defined set;

$\dfrac{\partial F}{\partial t} \geqslant 0$, where $t \in T$, and $T \subset \mathbb{R}$ occurs in the domain of F;

$\dfrac{d\,\mathsf{F}}{dt} = \mathsf{G}(\,\mathsf{F}, t)$, with G a specific function;

$\int_{t_1}^{t_2} F(q, \dot{q}, t)\, dt = \text{extremal}$, with t_1, t_2 two selected elements of $T \subseteq \mathbb{R}$;

$F_2(x, y) = \int_V du\ dv\ F_1(u-x, v-y)$, with $F_1, F_2 : E^3 \times E^3 \to \mathbb{C}$;

$\mathbf{v}^2 F_1 = F_2$, with $F_1, F_2 : E^3 \to \mathbb{R}$.

The preceding considerations suggest adopting the following characterization.

Definition 2. Let F be a state function for a thing. Any restriction on the possible values of the components of F and any relation between two or more components is called a *law statement* iff (i) it is included in a consistent factual theory and (ii) it has been confirmed satisfactorily (for the time being).

(It might be objected that, since we know that a function is a state function only once we have satisfied ourselves that it occurs in a law statement, the above definition is circular. Not so, because we have not *defined* state functions in terms of law statements: we have asserted only that lawfulness is a *criterion* or *test* of state functionality.)

If a law statement concerns a certain thing x, we may call it $L(x)$. (We shall presently see that this is not a matter of mere notation but one of conception.) And we may call $L(x)$ the totality of laws, or rather law statements, for thing x. Any member of this set may be construed as the value of a certain function L – a *law function* – with domain equal to the class of things concerned, and codomain the set of law statements of the form $L(x)$. For example, Ohm's law for a battery-resistor circuit x (an individual of a certain class C) can be written:

$$\text{For every } x \in C, \ L(x) = \ulcorner e(x) = R(x) \cdot i(x) \urcorner$$

where e, R and i are the electromotive force, the resistance, and the current intensity respectively. Hence in this case the predicate or (law) function L is the function $L: C \to L(C)$ such that, for any $x \in C$, $L(x)$ equals Ohm's expression. This manner of writing shows clearly that laws interrelate properties and are themselves properties of things. If a more explicit illustration of this claim is called for, one may write in all detail:

$$e: C \to \mathbb{R} \ , \ i: C \to \mathbb{R} \ , \ R: C \to \mathbb{R} \ , \text{ and } L: C \to L(C)$$

such that the above condition holds.

We can give an alternative rendering which lends itself better to generalization. Consider the set C of battery *cum* resistor circuits. As a state vector for things of this kind we may choose either the voltage–current pair or the voltage–charge pair. We pick the former:

$$F : C \xrightarrow{\langle e, \ i \rangle} \mathbb{R} \times \mathbb{R} \ .$$

This function assigns to each thing $x \in C$ the pair of real values $\langle e(x), i(x) \rangle$ in such a way that $e(x) = R \cdot i(x)$, where R is a real number. We now define the first and second projections, G and H, of F, as follows:

$$G: \mathbb{R} \times \mathbb{R} \to \mathbb{R} , \ H: \mathbb{R} \times \mathbb{R} \to \mathbb{R}$$

such that

$$\langle e(x), i(x) \rangle \xmapsto{G} e(x)$$
$$\langle e(x), i(x) \rangle \xmapsto{H} R \cdot i(x)$$

Since

$$(G \cdot F)(x) = e(x) \ \text{ and } \ (H \cdot F)(x) = R \cdot i(x),$$

Ohm's law now reads

$$G \cdot \mathsf{F} = H \cdot \mathsf{F}$$

i.e., as a statement of the identity of two function compositions.

The preceding treatment carries over to a situation where a fixed thing is being considered and the domain of the state function is taken to be some set T, such as the collection of all instants, or a portion of a certain manifold. We could multiply the examples, analyzing increasingly complex cases, but this would not lead us very far. It will be more profitable to look into a couple of very general frameworks for scientific theories, so general in fact that they cover all of the known basic physical theories.

5. LAGRANGIAN LAW SCHEMATA

One general framework for casting a variety of specific theories is of course Lagrangian dynamics. Consider first the case of global Lagrangian models of things, where spatial relations are of no concern. For the sake of perspicuity let us consider the simplest case, namely where the state function for the thing modelled has just two components, namely a time dependent generalized coordinate q, and a time dependent generalized velocity $\dot{q}$. ('Generalized' means of course that q need not be interpreted as a position coordinate, hence $\dot{q}$ need not be interpretable as a velocity proper. Actually not even t need be interpreted as time.) That is, let

$$\mathsf{F}: T \to \mathsf{R} \times \mathsf{R} \quad \text{with} \quad \mathsf{F}(t) = \langle q(t), \dot{q}(t) \rangle, \quad t \in T \subseteq \mathsf{R}.$$

Depending on the nature of the system, one manufactures a Lagrangian function

$$\mathscr{L}: \mathsf{R} \times \mathsf{R} \to \mathsf{R} \quad \text{with values} \quad \mathscr{L}(q(t), \dot{q}(t)).$$

This function is assumed to be first order differentiable in both arguments and subject to the law statement (the Euler–Lagrange equation)

$$\frac{\mathrm{d}}{\mathrm{d}t} \frac{\partial \mathscr{L}}{\partial \dot{q}} - \frac{\partial \mathscr{L}}{\partial q} = 0.$$

So far we have treated global (nonspatial) state functions. A system of fields, whether classical or quantal, calls for a slight extension of the Lagrangian formalism, which we shall presently indicate in outline. A suitable state function for a field system is

$$F = \langle F^\alpha, F^\alpha_{,\nu} \rangle, \quad \text{where} \quad F^\alpha : E^4 \to \mathbb{C}, F^\alpha_{,\nu} = \frac{\partial F^\alpha}{\partial x^\nu},$$

$\alpha = 1, 2, \ldots, n$, and $\nu = 0, 1, 2, 3$ ($x^0 = ct$, $x^1 = x$, $x^2 = y$, $x^3 = z$).

The index α labels either the different components of a single field, or the components of a system of superposed fields of a given kind (e.g., electromagnetic or meson fields). The state function F maps then space-time (E^4) into $\mathbb{C}^{5n}$. (Actually it is only a partial function, as it maps only a region of space-time.) Next, for each system one defines a Lagrangian density $\mathscr{L}$ whose values are obtained by composing F with $\mathscr{L}$.

We can see from the preceding that Lagrangian dynamics is an extremely general framework for couching specific scientific theories of almost any description. In fact the Lagrangian (or Lagrangian density, as the case may be) and the generalized coordinates and velocities upon which it depends may refer to almost any concrete system, whether physical or chemical, biological or social. Every interpretation of such functions, in terms of properties of things of some species of system, produces a specific scientific theory. This is why Lagrangian dynamics is found nowadays not only in mechanics but also in field theories, in biology and in management science – in short wherever change is describable in terms of differential equations.

As long as the generalized coordinates and velocities are not interpreted in factual terms, i.e. as properties of concrete things, one has a general scientific theory – so general in fact that it qualifies as an ontological theory, i.e. a theory concerned with things of any kind. This is perhaps the first time in the history of science that a theory born in a special science got generalized to the point of becoming a philosophical theory. But it is not the only case: automata theory is another such general framework. The existence of such extremely general theories, which are scientific as well as philosophic, suggests a new view on the relation between science and philosophy. In a nutshell it is this: Philosophy is general science, and science is special philosophy.

So much for the Lagrangian framework. As for the Hamiltonian one, it is of course just as important but we need not treat it separately here since in it, too, a law statement may be regarded as an equality of function compositions. What does merit special mention is the following point, which will come to prominence in the next section.

For each particular Lagrangian or Hamiltonian function (or operator) the final law statement, or a consequence (e.g., an integral) of it, will be a functional relation among the components of the state function. In the simplest case, where $F = \langle q, p \rangle$, one ends up with some family of functions which, written in implicit form, look like this: $G_\beta(q, p) = 0$, where β is some parameter (see Figure 1).

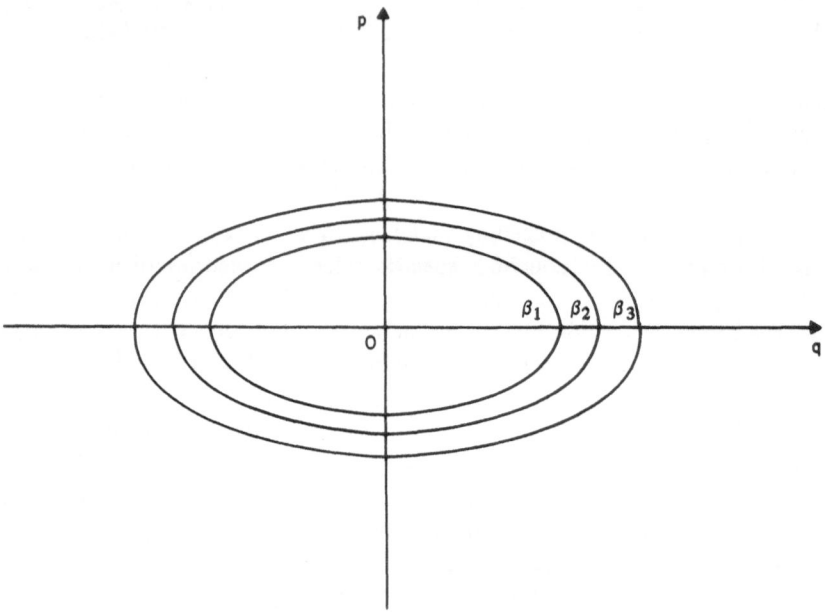

Fig. 1.   The state (or phase) space for a linear harmonic oscillator in the classical hamiltonian model. Here the parameter β that individuates the trajectories is the total energy of the system.

For each value of the parameter, the graph of the corresponding function in the coordinate-momentum plane represents the possible histories of the thing: one life line per value of β. The collection of all such possible histories is what may be called the state space of the thing. Needless to say the same holds, *mutatis mutandis*, for complex things or systems, in which case the state space has a larger number of dimensions. But the matter of state spaces deserves a separate section.

6. STATE SPACES

In every construction of a mathematical model of a thing the concern is with representing the possible (lawful) states, and perhaps also the possible (lawful) changes of state, of a thing. Hence the model is centered on a state space for the thing referred to. A few typical examples should give us a feel for this matter.

Example 1. In the elementary theory of the ideal gas, the state function is the triple consisting of the pressure, volume and temperature functions. The corresponding state space is a cube contained in $(\mathbb{R}^{+})^{3}$.

Example 2. In the genetics of populations, three state functions that are often employed are the size N of a population, the frequency (or rather probability) p of some particular gene, and the latter's adaptive value v. Hence for a system composed of two interacting populations, A and B, the state space is the region of $\mathbb{R}^{6}$ spanned by $\langle N_{A}(t), N_{B}(t), p_{A}(t), p_{B}(t), v_{A}(t), v_{B}(t) \rangle$ in the course of time.

Example 3. In the theory of social structure the instantaneous state of a community may be construed as the distribution of its population among the various social groups in the community. Here the component F_{i} of the total state function may be taken to be the column matrix $\|N_{ji}\|$ for a fixed i, the elements of which are the populations of the (mutually disjoint) social groups ensuing from the partition of the total population (at the given time) by the ith equivalence relation with a social significance (e.g., equal occupation or similar educational level).

Example 4. In chemical kinetics the instantaneous state of a chemical system is described by the values of the partial concentrations of both reactants and products. Therefore the state space of the system is inside $(\mathbb{R}^{+})^{n}$, where n is the number of system components (reactants, catalyzers and products).

Example 5. In elementary electrostatics the state function is $\mathbb{F} = \langle \rho, \varphi \rangle$, where ρ is the electric density and φ the electric potential. Hence the local state of the given field is the value of $\mathbb{F}$ at $x \in E^{3}$, and the entire state space is the set of ordered couples $\{\langle \rho(x), \varphi(x) \rangle \in \mathbb{R}^{2} \mid x \in V \subseteq E^{3}\}$, where V is the spatial region occupied by the field. For the system of n fields (or n components of a single field) treated in the preceding section, the state of the system at $x \in E^{4}$ is $\mathbb{F}(x) = \langle F^{\alpha}(x), F^{\alpha}_{,\nu}(x) \rangle$, which is a $5n$-tuple of complex numbers. Hence the state space of the system is included in $\mathbb{C}^{5n}$.

Example 6. In quantum mechanics the state of a system is represented by a one dimensional subspace (or ray) of the Hilbert space associated with the system. Since a thing of this kind typically is not attributed a point-like location but is assumed instead to be spread over some spatial region $V \subseteq E^3$ (with a definite probability distribution), the state of the thing is the set of all values its state vector takes in V.

Before attempting to draw general conclusions let us emphasize a point of method made in Section 3. We may certainly assume that, whether we know it or not, each (isolated) thing – in particular each closed system – is in a definite state relative to some reference frame and at each instant (or else point in space-time). Yet our *representation* of such a state will depend upon the state function chosen to represent the thing – which choice depends in turn upon the state of our knowledge as well as upon our goals. What holds for each single state holds *a fortiori* for the entire state space for a thing. That is, far from being something out there, like physical space, a state space for a thing stands with one leg on the thing, another on a reference frame, and a third on the theoretician (or modeller). To persuade oneself that this is so, suffice it to take a new look at Example 5 above, where a reference frame at rest relative to the field source was assumed. If now the system is considered relative to a moving frame, a four-vector current density will have to replace the single charge density, and a four-vector potential will take the place of the single scalar potential. Alternatively – and this is where the scientist's freedom comes in – the four potential can be replaced by an antisymmetric tensor representing the electric and magnetic components of the field relative to a given frame.

Having emphasized the conventional ingredient of every representation of states, let us now stress that every such representation has an objective basis as well. For one thing a state function may not take values in its entire codomain but may be restricted to a subset of the latter, and this by virtue of some law. (Recall Section 4.) Therefore, for every component of a state function for a thing, the focus of our concern will be the range of it rather than its codomain. If we now take the whole set of components for a given thing and form the Cartesian product of their respective ranges (in tune with Definition 1), we obtain the *conceivable state space* for that thing. This is precisely what we did in the examples that led off this section.

However, this restriction is insufficient, as should become apparent when we consider the following examples. The total population of organisms of a given kind in a given territory is constrained not only by the carrying capacity of the latter but also by the birth and death rates, as well as by additional factors such as sunshine and rainfall. Again, although the range of the speed function for a body is the entire real interval $[0, c)$, the speed of an electron traveling in a transparent medium will not come close to the upper bound c, for such a body is subject to further laws. In general: Only those values of the components of the state function that are compatible with the laws will be really (not just conceptually) possible. In other words, because the laws impose restrictions upon the state functions and their values, hence upon the state spaces, only the states in certain subsets of the latter are accessible to the thing. We shall call the accessible part of a state space the *lawful state space* of the thing (in the given representation and relative to a given frame). To say that a thing behaves lawfully amounts then to saying that the point representing its (instantaneous) state cannot wander beyond the bounds of the state space chosen for the thing. See Figure 2.

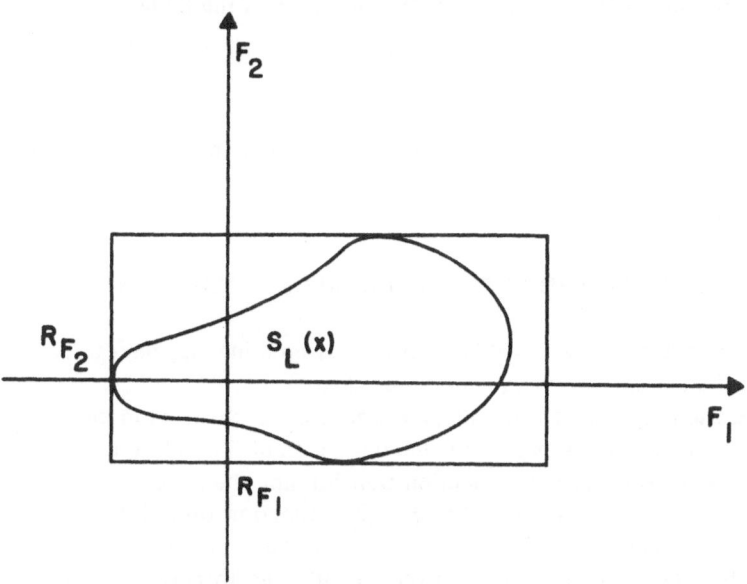

Fig. 2. The lawful state space for a thing is a subset of the Cartesian product of the ranges of the components of the state function (in this case only two).

In terms of the concepts introduced in Section 4, the preceding remarks can be summarized as follows:

Definition 3. Let $F = \langle F_1, F_2, ..., F_n \rangle\colon A \to V_1 \times V_2 \times ... \times V_n$ be a state function in a model for a certain thing x and call $\mathbb{L}(x)$ the set of all law statements of x. Then the subset of the codomain $V_1 \times V_2 \times ... \times V_n$ of F restricted by the conditions (law statements) in $\mathbb{L}(x)$ is called the *lawful state space* of x, or $S_L(x)$ for short:

$$S_L(x) = \{\langle x_1, x_2, \ldots, x_n \rangle \in V_1 \times V_2 \times \ldots \times V_n \mid F \text{ satisfies } L(x)\}.$$

Clearly, the lawful state space is included in the corresponding conceivable state space.

Example. In Example 5 above, the conceivable state space of an electrostatic field f was

$$S(f) = \{\langle \varphi(x), \varphi(x) \rangle \in R^2 \mid x \in V \subseteq E^3\}.$$

Since the two components of the state function $F = \langle \varphi, \varphi \rangle$ are linked by the law statement: $\nabla^2 \varphi = 4\pi\rho$, the lawful state space of the thing is

$$S_L(f) = \{\langle \rho(x), \varphi(x) \rangle \in R^2 \mid x \in V \subseteq E^3 \ \& \ \nabla^2 \varphi(x) = 4\pi\rho(x)\}$$

and $\quad S_L(f) \subseteq S(f).$

Note that, because of the restrictions imposed by the law statements, lawful state spaces, though usually included in vector spaces, need not be vector spaces themselves.

7. LAW STATEMENTS AND TRANSFORMATION FORMULAS

Recall now that every law statement for a given thing may be regarded as the value of a certain function L which we called the corresponding law function. (Cf. Section 4.) For a fixed thing, the arguments of the law function are components of the state function or functions thereof. Therefore a law function may be construed also as a function transforming the state space into itself. That is, for any given thing, $L: S \rightarrow S$ is a function on S that assigns each state $s \in S_L$ another state $L(s) \in S_L$, not of course an arbitrary one but one lawfully related to the former. Consequently the totality L of law statements for a given thing is contained in the set of all the transformations of its state space. This point will serve as a basis for further developments in Section 7 on change.

Not every transformation of a state space is a law statement. In fact certain transformations of S_L, though lawful (i.e., law preserving), do not represent laws but just choices of different state functions. Consider a state function transformation $F^* = f(F)$ for a given thing, where f is a function subject to certain restrictions. Every such transformation ensues in a different representation of the states of the thing, i.e., in a different state space S'. In other words, if $f: S_L \rightarrow S'_L$ is such a bijection, then $f(S_L)$ is an alternative state space for the same thing. (See Figure 3.)

But of course not all such representation changes are admissible: some will

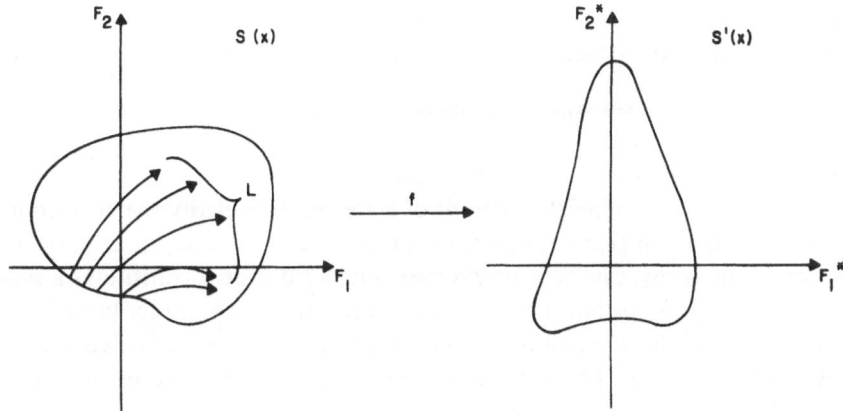

Fig. 3. The laws of thing x are transformations of $S(x)$ into itself, e.g., $L = \lceil pv = \text{const.}\rceil$ sends point $\langle p, v \rangle$ into point $\langle p', v' \rangle$ such that $pv = p'v' = \text{const}$. On the other hand the transformation formulas (such as the canonical ones) link two different representations, S and S', of the states of the system.

keep the laws L invariant while others will violate them. We consecrate this important dichotomy by introducing:

Definition 4. Let $F: A \rightarrow V$ be a state vector for a given thing x. Then:

(i) the *lawful transformations* of F are those transformations $f: V \rightarrow V'$ such that $F* = f \circ F$ and the laws L of x are left form-invariant;

(ii) every lawfully transformed state space $f(S_L)$ is an *equivalent representation* of the states of x;

(iii) the collection all equivalent representations of the states of x constitutes *the representation* of the states of x with the help of the given state functions.

Example 1. In special relativistic theories the position and time coordinates transform according to the Lorentz formulas. These transformations induce transformations of any state function, such that the basic equations retain their form, or are Lorentz-*covariant*. Those state functions that do not change are called *invariant*.

Example 2. In Hamiltonian theories the state variables are the generalized coordinates q and momenta p (one set of each for every component of the system). These variables span the state (or phase) space of the system. New state functions $F* = \langle f(q, p), g(q, p) \rangle$ can always be introduced. But only those will be admissible which leave the canonical equations invariant, i.e., those for which $q^* = f(q, p)$ and $p^* = g(q, p)$ are such that

$$\dot{q}^* = \partial H^*/\partial p^* \quad \text{and} \quad \dot{p}^* = -\partial H^*/\partial q^*.$$

Such transformations are called *canonical*.

Finally let us employ the concept of state space to clarify the distinction between a thing in general and a system in particular. An analysis of the state space of an entity can help us discover whether or not it constitutes a system, i.e., whether or not it has parts that act upon one another. Indeed the state space of an aggregate of non-interacting things is uniquely determined by the partial state spaces. Moreover, since the contributions of the latter are all on the same footing, we may take the total state space to equal the union of the partial state spaces. Not so in the case of a system: here the state of every component is determined, at least partly, by the states other components are in, so that the total state space is no longer the union of the partial state spaces. An example or two will help bring this point home.

Think of the state spaces of a moth and a candle before and after the former spirals into the latter. If a more dignified illustration is wished, think of a system composed of two electrons close enough to interact appreciably, as in the case of those of a helium atom. This system is described by the corresponding Schrödinger equation (classically, by two coupled equations of motion) jointly with the Pauli exclusion principle. The latter selects those state functions ψ that are odd or antisymmetric in the coordinates of the electrons. (That is, the additional law is $\psi(x, y) = -\psi(y, x)$, where x and y are the coordinates of the electrons. It follows that $\psi(x, x) = 0$, i.e. the two particles cannot coexist at the same point – unless they have different spins.) This principle expresses a global or systemic property, one that the individual components fail to possess, whence it cannot be represented in the partial state spaces. Therefore the construction of the state space of the system must proceed from scratch rather than on the sole basis of the state spaces of the individual electrons.

We summarize and generalize the preceding remarks in the following postulate. Let $S_f(x)$ and $S_f(y)$ be lawful state spaces of things x and y respectively, both relative to the same frame f. (We refrain from using the subindices L_1 and L_2 for simplicity.) Further, call $z = x \dotplus y$ the thing composed of x and y. Then, for all reference frames $f \in F$, $S_f(z) = S_f(x) \cup S_f(y)$ iff x and y do not act upon each other.

(Actually the mere wording of this axiom presupposes the notion of association or juxtaposition of things, and of action of one thing upon another, which we have not introduced in the present work.)

The preceding assumption allows us to propose the convention we are after:

Definition 5. Let x be a thing composed of things x_i for $1 \leqslant i \leqslant n$. Then x is an *aggregate* (or *conglomerate*) if and only if, for all choices of state functions, the state space $S(x)$ equals the union of the partial state spaces $S(x_i)$ for $1 \leqslant i \leqslant n$. Otherwise x is a *system*. [And, whether it is an aggregate or a system, the parts x_i of x are called its *components.*]

Obvious examples of systems are molecules and even every atomic component of a molecule. On the other hand the nucleons and electrons composing the atoms are not systems since they have no separate components. (The distinction between the particle part and the field part of an electron or some other charged entity is artificial. Physical theory treats the thing as a whole without components.) Another non-example of a system is the aggregate formed by a body and a reference frame. Since reference frames are supposed to record without either being influenced or influencing back, a state space for the aggregate must be taken to be the union of the partial state spaces.

8. EVENTS AND PROCESSES

Every event occurs in or to some concrete thing, and it consists in a change of state of the thing – the change being merely quantitative, as in the case of motion, or qualitative as well, as in the case of the coming into being or the disappearance of a thing. A light flash, the dissociation of a molecule, a storm, a chemical reaction, the growth of a bud, the learning of a trick, and the fall of a government are events. These particular events may also be called processes, for – unlike the instantaneous collision of two atoms – they are complex, so they can be analyzed into further events.

Being changes of state of things, events and processes are representable as trajectories in the state spaces of changing things. And because states are relative to the reference frame and the representation (in particular the choice of state functions), their changes too are relative in the same sense. Hence one may well contrive a state space in which the representative point is stationary; but the point is bound to move in most other state spaces. Change, then, though relative – to frame and representation – is not necessarily subjective or illusory.

Different trajectories in a state space may have the same end points. In other words, there are cases in which one and the same net change can be effected along alternative routes. See Figure 4.

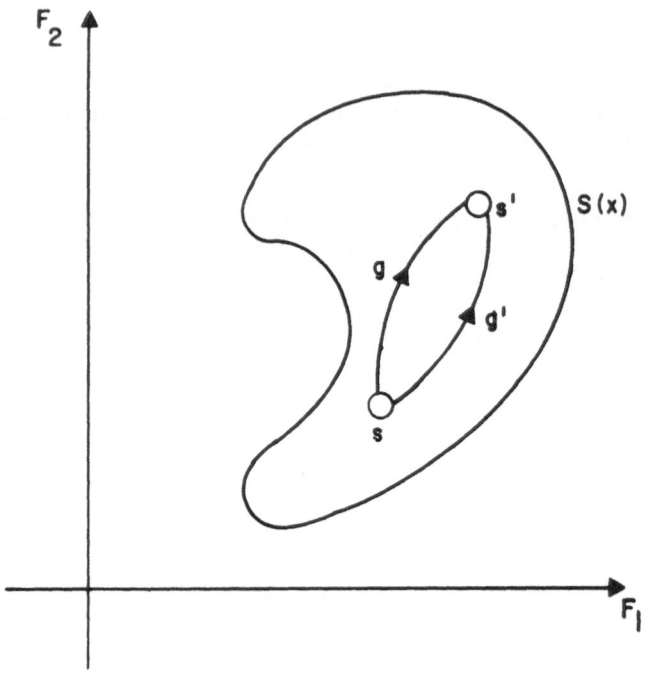

Fig. 4. Two different processes resulting in the same net change. The net change from state s in state space $S(x)$ to state s' in the same state space can be represented as the ordered pair $\langle s, s' \rangle$. But since the change along curve g may be distinct from the change along curve $g' \neq g$, we must represent the full events (or processes) by $\langle s, s', g \rangle$ and $\langle s, s', g' \rangle$ respectively.

Now, the functions g and g' occurring in our previous illustration are not supposed to be arbitrary: they must be lawful if we are to allow only lawful events and discard lawless ones, i.e., miracles. In other words, g and g' must be compatible with the laws of the thing. (They need not be law functions: in general they will represent laws *cum* circumstances, e.g., field laws together with specifications of the field sources and boundary conditions.) This suggests introducing:

Definition 6. Let $S(x)$ be a state space of thing x, and let $s, s' \in S(x)$ be two states of x relative to $S(x)$. Then a *lawful event* (or *process*) with end points s and s' will be representable by a triple $\langle s, s', g \rangle$, where $g: S(x) \to S(x)$ is compatible with the laws of x.

If we now fix the transformation g, i.e., if we focus on events of a certain kind (e.g. mechanical or genetic), and discard the intermediate states between the given end points of the net change, we are left with ordered pairs $\langle s, s' \rangle \in S_g(x) \times S_g(x)$. The collection of all such pairs of states for a given g constitutes then the *space of events* of x for g. Symbol: $E_g(x) \subseteq S_g(x) \times S_g(x)$. In general the inclusion is proper precisely because g will exclude certain conceivable but physically impossible changes.

9. EVENT SPACE

We shall begin our investigation of the structure of event spaces by examining the simple case of a three state thing – or rather a thing that, in the respect of interest, can be in either of three states – e.g. on, off, and transient, as in the case of a switch. Calling these states a, b and c, we have $S(x) = \{a, b, c\}$. We now form the ordered pairs:

$\langle a, a \rangle$ (Thing x stays in state a, i.e., the identity event at a.)

$\langle a, b \rangle$ (Thing x goes from state a to state b.)

etc. We have thus altogether nine elementary (not composite) events in $S(x) \times S(x)$:

$$e_1 = \langle a, a \rangle \equiv u_a, \quad e_4 = \langle a, b \rangle, \quad e_7 = \langle b, a \rangle$$
$$e_2 = \langle b, b \rangle \equiv u_b, \quad e_5 = \langle b, c \rangle, \quad e_8 = \langle c, b \rangle$$
$$e_3 = \langle c, c \rangle \equiv u_c, \quad e_6 = \langle a, c \rangle, \quad e_9 = \langle c, a \rangle.$$

That is, $(S(x))^2 = \{u_a, u_b, u_c, e_4, e_5, e_6, e_7, e_8, e_9\}$. Suppose now, for the sake of definiteness, that all these events are lawful, i.e., really possible. That is, set $E_f(x) = S_f(x) \times S_f(x)$. A standard representation of this event space is its Moore graph (Figure 5).

We assume next that the composition of certain events is possible. For example, event $e_4 = \langle a, b \rangle$ can be followed by event $e_5 = \langle b, c \rangle$, and this complex event may be regarded as a decomposition of event $e_6 = \langle a, c \rangle$. To symbolize the composition of events we use the asterisk and write: $e_4 * e_5 = e_6$ or, explicitly, $\langle a, b \rangle * \langle b, c \rangle = \langle a, c \rangle$. (Here intermediary states do not show up in the net change: they are absorbed.) On the other hand the complex event $\langle b, c \rangle * \langle a, b \rangle$, which is the reverse of the former, does not occur, and so it is left undefined. In other words, to represent the combina-

MARIO BUNGE

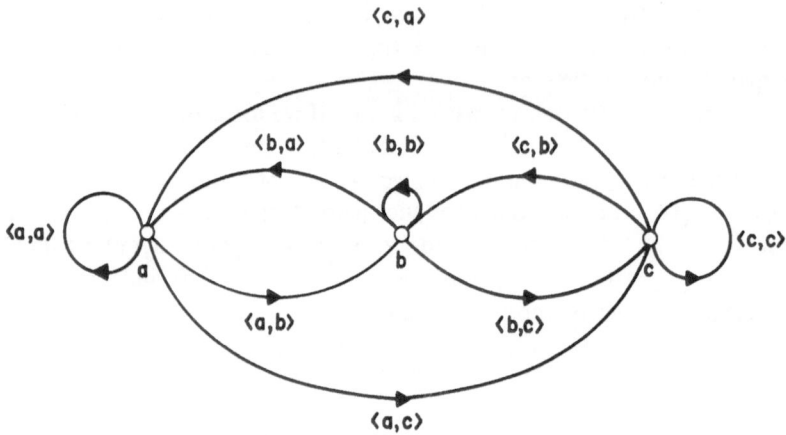

Fig. 5. The category of events conceivable for a three state thing.

tion of events we introduce a partial binary operation $*$ in $E(x)$. If e and f are in $E(x)$, then $e * f = g$ is another event in $E(x)$ consisting of event e *followed by event f.* (In some cases the composition is not defined, i.e., $*$ is a partial operation.) The possible binary combinations of events occurring in a three state thing are shown in the following table. Each entry in this multiplication table shows the net change, not the change process. But of course the table allows one to analyze certain events as processes, i.e. sequences of elementary events.

$*$	$\langle a, a \rangle$	$\langle a, b \rangle$	$\langle a, c \rangle$	$\langle b, a \rangle$	$\langle b, b \rangle$	$\langle b, c \rangle$	$\langle c, a \rangle$	$\langle c, b \rangle$	$\langle c, c \rangle$
$\langle a, a \rangle$	$\langle a, a \rangle$	$\langle a, b \rangle$	$\langle a, c \rangle$						
$\langle a, b \rangle$				$\langle a, a \rangle$	$\langle a, b \rangle$	$\langle a, c \rangle$			
$\langle a, c \rangle$							$\langle a, a \rangle$	$\langle a, b \rangle$	$\langle a, c \rangle$
$\langle b, a \rangle$	$\langle b, a \rangle$	$\langle b, b \rangle$	$\langle b, c \rangle$						
$\langle b, b \rangle$				$\langle b, a \rangle$	$\langle b, b \rangle$	$\langle b, c \rangle$			
$\langle b, c \rangle$							$\langle b, a \rangle$	$\langle b, b \rangle$	$\langle b, c \rangle$
$\langle c, a \rangle$	$\langle c, a \rangle$	$\langle c, b \rangle$	$\langle c, c \rangle$						
$\langle c, b \rangle$				$\langle c, a \rangle$	$\langle c, b \rangle$	$\langle c, c \rangle$			
$\langle c, c \rangle$							$\langle c, a \rangle$	$\langle c, b \rangle$	$\langle c, c \rangle$

10. THE CATEGORY OF EVENTS

A number of regularities emerge clearly from the table in the last section. We make bold and assume that they hold for every event space and summarize them in:

Definition 7. Let $S(x) \neq \emptyset$ be state space for a thing x and let $E(x) = S(x) \times S(x)$. The triple $\chi = \langle S(x), E(x), * \rangle$, where $*$ is a partial (not everywhere defined) binary operation in $E(x)$ such that, for all a, b, c, d in $S(x)$,

$$\langle a, b \rangle * \langle c, d \rangle = \begin{cases} \langle a, d \rangle \text{ iff } b = c \\ \text{not defined if } b \neq c \end{cases}$$

is the *event space* of x associated with $S(x)$ iff

(i) every element of $E(x)$ represents a conceivable change of (or event in) x;

(ii) for any $e, f \in E(x)$, $e * f$ represents the event consisting of event f following event e;

(iii) for any $s \in S(x)$, $\langle s, s \rangle \in E(x)$ represents the identity event (or non-event) at s, i.e., the staying of x in state s.

A immediate consequence is that every conceivable event $\langle s, s' \rangle$, where $s, s' \in S(x)$, has a unique conceivable converse, namely $\langle s', s \rangle$. (Which goes to show that $E(x)$ is the totality of conceivable events, not just of really possible ones.) Another consequence is:

Corollary 1. For every $s \in S(x)$ there is an identity $i_s = \langle s, s \rangle$ such that, for each $e \in E(x)$ for which $*$ is defined, $e * i_s = i_s * e = e$.

In other words, the event space $\chi = \langle S(x), E(x), * \rangle$ is a *category* with set of objects $S(x)$, set of morphisms $E(x)$, and identity morphisms i_s, for all $s \in S(x)$; the composition is $*$. The transition graph in Figure 5 is a vivid pictorial representation of this category in the case where $S(x)$ has only three members.

Note that in the category χ the following holds: For any two items (states) $s, s' \in S(x)$ there is exactly one morphism $s \mapsto s'$, namely $\langle s, s' \rangle$ This in turn implies that every morphism in the category is an isomorphism, i.e., the inverse of $\langle s, s' \rangle$ is the unique morphism $s' \mapsto s$. This entails:

Corollary 2. Let $s, s' \in S(x)$ be states of x. Then

(i) no event other than an identity event is immediately repeatable: $\langle s, s' \rangle * \langle s, s' \rangle$ is not defined in $E(x)$;

(ii) if an event is followed by its converse, no net change results:

$$\langle s, s' \rangle * \langle s', s \rangle = \langle s, s \rangle.$$

Hence for an event to be repeated the thing must first go back to the initial state, either through the converse event or through intermediate events, as in the case where $\langle s, s' \rangle * \langle s', s'' \rangle * \langle s'', s \rangle = \langle s, s \rangle$.

From Definition 7 it follows that intermediary states play no role in the net result. Thus the processes

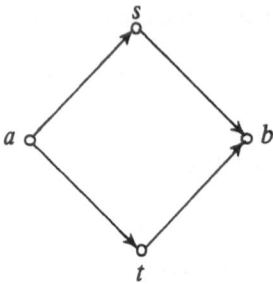

are equivalent in that $\langle a, s \rangle * \langle s, b \rangle = \langle a, t \rangle * \langle t, b \rangle = \langle a, b \rangle$. Therefore we can make:

Definition 8. Complex events in a given event space are *equivalent* iff they have the same outcome [i.e. iff they relate the same initial and final states].

The net event approach is of course eminently suitable for comparing, say, brains to computers, since in the case of either only the net results are of interest – not how and why one thing goes from one state to another. Such limited concerns aside, the study of process must focus on the entire triples $\langle s, s', f \rangle$ introduced in Definition 6.

11. CONCLUDING REMARKS

We have tried to clarify a few notions that play a central role in the general theories of systems, in science, and in philosophy – mainly those of state, law, and event. Our definitions and analyses do not constitute a theory, for very little follows from them. They form instead a framework that, enriched with stronger assumptions, may suggest or shelter a number of theories. Because of this our paper belongs in the foundations and the philosophy of systems theory rather than being a regular member of the latter. Indeed, our topics have been extremely general, therefore our assumptions mild to the point of vacuousness, and in any event they aim at clarifying certain notions rather than at describing the actual behavior of particular systems. Perhaps the sole interest of this paper is in its showing that the notion of a state space has applications everywhere instead of being confined to automata theory and the theory of linear mechanical or electrical systems. And, since it is so pervasive a concept, that of state has a good chance of becoming a key concept in scientific ontology. What holds for the concept of a state holds also for that of an event, which has heretofore been either conceived in rather narrow (specific) ways or misidentified with the location of an event in spacetime, or even made the stuff things are made of.

ACKNOWLEDGEMENTS

I take pleasure in thanking my former research associates Professors William E. Hartnett (SUNY at Plattsburgh) and Arturo A. L. Sangalli (University of Ottawa) for many a pleasant and enlightening discussion during the preparation of this paper. Also I am grateful to the Canada Council for a research grant that made the above-mentioned discussions possible.

CHAPTER 6

UNDERSTANDING SOCIAL AND ECONOMIC CHANGE IN THE UNITED STATES*

JAY W. FORRESTER

I had originally intended to deal with the philosophy of modeling which underlies (Forrester, 1961, 1969, 1971b) and about ideas growing from various critiques (Oltmans, 1974; Forrester, Low and Mass, 1974) of these books. That talk would have been about the need for a more penetrating treatment of the philosophy and theory underlying simulation models, and such issues as aggregation, the use of data for establishing confidence in models, and the boundaries of models (Forrester, 1973). But I have been dissuaded from that subject by your program committee. Instead, a description has been requested of the model of social and economic change in the United States on which we are now working at MIT.

We are well advanced on a national model of social and economic behavior. It is a system dynamics model and so is very different from the more common econometric models. The present controversies about the economy, uncertainties about the causes of inflation, and debates about economic theory all suggest the need for a new approach to economic dynamics. We believe there is an excellent chance that a comprehensive system dynamics model incorporating the structures that generate economic fluctuations, growth, and environmental restraints can complement other approaches and can fill in where other methods of analysis have been unable to answer important questions.

The system dynamics model now nearing completion should yield substantial new understanding of the major social and economic pressures confronting the United States. Within the next few months the model will be far enough along for us to start examining the forces underlying inflation, the nature of the new economic mode that can simultaneously produce inflation and unemployment, the impact on standard of living as the United States buys more energy and resources from abroad, the consequences of

---

* Revised and slightly edited version of a paper presented to the Summer Computer Simulation Conference, Houston, Texas (1974).

W. E. Hartnett (ed.), Systems: Approaches, Theories, Applications, 97–119.

various methods of recycling money paid for oil imports, the effect on exchange rates from foreign manufacturing by multi-national corporations, and the economic forces arising to reverse the historical flow of people from agriculture and manufacturing toward government services.

The inadequacy of past precedents and rules of thumb are becoming evident. Only through modeling will we be able to develop more promising policies for the future, and to communicate new insights widely enough to establish public support for measures necessary to cope with an increasingly difficult future.

Development of the national model at MIT has been under way about two and a half years with financial sponsorship from the Rockefeller Brothers Fund. The model is much more extensive in scope and depth than the earlier work. The *Urban Dynamics* and *World Dynamics* books were both unsponsored efforts with no outside financial support. Without a research staff, the scope was sharply curtailed. *Urban Dynamics* grew out of a four-month collaborative effort between John Collins, former Mayor of Boston, and myself. As some of you know, the *World Dynamics* model was created as a discussion vehicle for a conference on world issues. Those projects were exploratory. The models were highly aggregated. The books have been criticized for short-comings that arose from limited time and budget. But neither simulation modeling in general, nor the system dynamics approach in particular, are intrinsically so limited.

The present national model is far more comprehensive than the earlier efforts. An able staff has been assembled. We took about a year to ponder objectives and structure. The last year and a half has been devoted to writing equations and developing sub-structures of the model. The pieces are now being assembled. Assembly will proceed in stages with intermediate testing of partial assemblies extending over several months.

1. SYSTEM DYNAMICS

The economic model now being assembled is a system dynamics model. As such it differs from more traditional economic models in structure, sources of input information, nature of validity testing, and purpose. A brief description of the system dynamics approach should help in understanding the model.

System dynamics is a way of combining personal experience with computer simulation to yield a better understanding of social systems. The field of system dynamics has been under development at MIT and elsewhere since 1956. Twenty or more books have been published on subjects ranging from

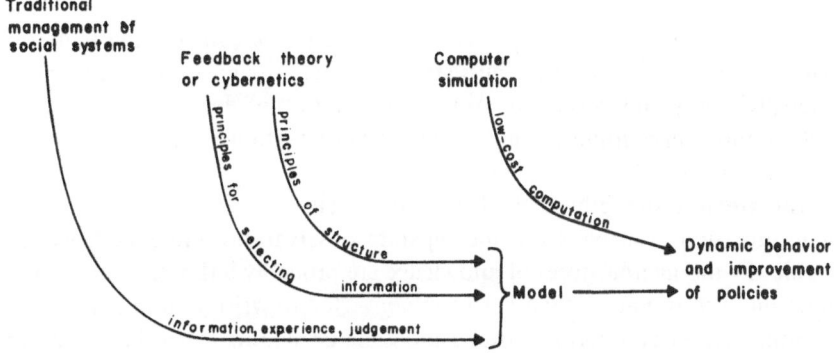

Fig. 1.   Background of system dynamics.

corporate policy to major world interactions. The principal books from the research programs have each been adopted as texts in dozens of universities.

1.1. *Background of System Dynamics*

System dynamics is perhaps best described in terms of the background threads on which it builds. In Figure 1, three earlier developments — traditional management of social systems, feedback theory, and computer simulation — combine to become system dynamics. Traditional management is the process used to govern social systems throughout history. Feedback theory or cybernetics is a body of methods and principles developed during the last hundred years dealing with how decisions, and the way they are imbedded in information channels, cause the dynamic behavior of systems. Computer simulation allows one to determine the time-varying behavior implicit in the complex structure of a system.

People start the traditional management process by observing the world about them, noting the pressures and reactions of people and groups, and detecting the linkages and flows of information and influence. From these observations people form mental images of the structure of a social system. From the mental images they attempt to anticipate what will happen next and how a different policy might make the system behave more desirably.

Traditional management processes guide our personal lives, family affairs, cities, countries, and international relations. It is the nearly universal approach to directing human activity. Because it is the basis of civilization and because it has served well, no quick or radical break with past tradition is either possible or desirable.

Traditional management, based on observation and judgement, has great strengths, but it also has serious weaknesses. Any new contribution to better management of social systems must start from the present practices and move gradually toward improvement. Any better method of decision making must build without discontinuity on the strengths of traditional management while compensating for the weaknesses.

The greatest strength of traditional management comes from the wealth of information available from the separate observations and experiences of people. In the mental stores of knowledge are probably a thousand or million times more information than has been converted to written form in libraries. In turn, written descriptions cover a thousand or million times the scope and richness of information that is available in measured and numerical form. If we are to improve on social decisions, we must be able to build on the most comprehensive information base available — the observations, knowledge, and judgement stored in people's heads. System dynamics uses that descriptive information along with any available written and numerical information.

But traditional management has several serious weaknesses. System dynamics helps to alleviate those weaknesses.

The first weakness in traditional management arises from the very wealth of information that is the greatest strength. In fact, we have too much information. We are flooded and overwhelmed with information. The traditional processes contain no general principles or organized philosophy for picking the relevant from the extraneous information. As indicated in Figure 1, principles drawn from feedback theory assist in choosing from the excess of information that relevant to the behavior modes of interest.

The second weakness of traditional management arises from lack of organizing principles for the structuring of information. Even if the first weakness is overcome and the relevant information and relationship are chosen, no guidelines exist for organizing the chosen assumptions into a structure that explains the observed system behavior. Again, feedback theory offers principles (Forrester, 1971a) for simplifying and organizing the structure of a system.

But even if information is effectively selected and usefully organized into a relevant model, traditional management encounters a third weakness. Even when assumptions are explicitly stated, the human mind is not well adapted to determining the future time-varying consequences of those assumptions. Different people, even when they accept the same assumptions and structure, often draw contrary conclusions. A consensus is hard to reach, and even a majority opinion may be incorrect. As suggested in Figure 1, computer simulation can be used to determine, without doubt, the future dynamic implications of a specific set of assumptions.

To summarize, system dynamics starts from the practical world of normal economic and political management. It does not begin with abstract theory nor is it restricted to the limited information available in numerical form. Instead it uses the descriptive knowledge of the operating arena about structure, along with available experience about the decision-making. Such inputs are augmented where possible by written description, theory, and numerical data. Feedback theory is used as a guide for selecting and filtering information to yield the structure and numerical values for a computer simulation model. Because the resulting models are too complex for either intuitive or mathematical solution, a computer simulates, or plays the roles, of the many participants in the system to determine how they interact with one another to produce changing patterns of behavior.

1.2. *Necessity of Models*

Models are not new in social decision making. System dynamics does not for the first time introduce models into the social and political process. Models have always been the basis for the traditional management methods.

Every decision we make is based on a model. One does not have a family, city, or nation in his head. He has only images, relationships, and abstractions from real life. These perceptions are models in the same sense that the word is used in system dynamics. One uses observation to form a mental image, or model. The mental model becomes the basis for decisions.

System dynamics does not impose models but is a way of improving on the models that would otherwise be used to manage human affairs. The system dynamics model is more explicit than a mental model, so it can be communicated with less ambiguity. A system dynamics model is more carefully structured in accordance with dynamic principles, so it better relates underlying assumptions to system behavior. A system dynamics model can be simulated on a computer, so, unlike a mental model, its behavioral implications can be determined precisely.

1.3. *Comparing System Dynamics with Econometrics*

System dynamics is a departure from conventional methodology in economic modeling. Most major economic models have used econometric methods to convert historical numerical time-series data into parameters for an assumed structure of equations. But such methods have failed to answer pressing questions about fundamental behavior arising from social, economic, and environmental interactions. The current national issues are so critical that a new approach should be tried. Compared to an econometric model, a system dynamics model:

(a) makes greater use of descriptive information and managerial and political experience;

(b) incorporates a broader range of variables and encompasses the many relevant disciplines outside of economics;

(c) uses numerical data from real life in a different way – in model construction to complement descriptive information, in validation to compare with corresponding output data from the model;

(d) generates social and economic fluctuations and growth from the internal feedback structure without using exogenous variables to drive change;

(e) includes important social and psychological variables for which statistical data are not available;

(f) explains how 'structural changes' occur as the economy moves into new modes of behavior that are not represented in past time-series data;

(g) facilitates incorporating the wide range of nonlinear structures that generate so much of observed real-world behavior;

(h) emphasizes the conservation of flows by including the buffer stocks that decouple the instantaneous flow rates;

(i) distinguishes more sharply between real variables, their money value, and information about them to capture the dynamic interactions between the real, money, and information aspects of a system;

(j) encourages constructions of a deeper substructure of feedback loops to represent causal mechanisms underlying macro-economic behavior;

(k) organizes structure so that each parameter has independent real-life meaning in the operating world and can be individually drawn from and checked against descriptive and quantitative information available at the place in the real system to which the parameter applies;

(l) serves as a more effective communication medium for resolving disagreements because of the way both model structure and parameters correspond with descriptive knowledge in the operating world;

(m) places more emphasis on the importance of internal structure;

(n) focuses more on understanding the reasons for observed behavior and on developing policies to produce better behavior, and focuses less on prediction;

(o) combines over a greater time span the short-term with long-range human objectives;

(p) permits a wider diversity of contact between the model and the real world to make validation more persuasive.

1.4. *Avoiding the Practical Difficulties with Statistical Models*

A system dynamics model can circumvent the major practical shortcomings that keep econometric methods from deducing correct parameter values and from correctly evaluating the validity of hypothesized structures. Dissatisfaction with econometric models is widely reflected in the economics literature. Although failures have usually been attributed to inadequate data, a more fundamental reason is emerging for the deficiencies in econometric models.

The statistical methodologies are based on precise theoretical foundations regarding correspondence between the hypothesized model structure and the structure of the real system, the nature of random disturbances in the real system, the characteristics of auto and cross correlation, the frequency of sampling of collected data, and the absence of measurement errors in the data. As recognized by almost everyone, none of the underlying theoretical requirements are fully met, so the practical application of econometrics rests on the assumption that the theoretical requirements are approached sufficiently closely for the methods to be applicable.

As long as econometric methods are applied only in real situations, no final test of methodological accuracy is possible because the true values of the real system parameters are not available for comparison with parameter values that have been deduced from statistical analysis. In ordinary practice the estimates of validity are entirely internal to the statistical process itself, so the estimate of validity also rests on the assumption that underlying theoretical requirements are adequately met.

But how closely must the demands of theory be approximated? The answer cannot be found in the real-life setting but can be determined in controlled laboratory experiments. A dynamic, feedback model can be *defined* as the 'real' system and used to generate, without exogenous driving variables, time-series data comparable to that collected in real situations. The *data-generation model* stands in the place of the real system. Separately, an *estimation model* is used by a statistical analyst as an approximation to the 'real' system as is done in the usual data-analysis procedures. All conditions of the process are completely controllable. The statistical analyst can be allowed to know as much or as little about the structure of the 'real' system as desired. The random noise processes within the 'real' system can be controlled. The interval between data samples can be independent of the dynamic processes within the 'real' system. Controllable errors can be inserted in the data collected from the 'real' system. The degree of correspondence between the structure of the 'real' system and the structure of the estimation model can be controlled. And finally, perfect knowledge is available about the structure and parameters of the 'real' system to permit a definitive evaluation of the parameters obtained by the statistical methods.

Such comprehensive examination of the practical aspects of statistical model-building has recently been started. The first tests, done on single-equation least squares regression analysis, reveal the consequences of various deviations from the theoretical assumptions underlying the standard procedures. It is beginning to appear (Senge, 1974) that least squares regression, which is probably the statistical method most often used by social scientists, is highly sensitive to surprisingly small departures from the idealized theoretical foundations. In fact, the statistical methods break down and become misleading with such minor departures from perfection that meeting the theoretical requirements closely enough seems inconceivable.

The laboratory tests indicate that the generalized least squares data analysis can give not only major errors in the estimates of parameters but also misleading indications from the internal validity measures. Accurate parameters can be obtained in a correct structure for the estimation model along with validity measures that suggest low confidence; depending on such results of the analysis would lead one to discard correct parameters and structure. Or, at other times, inaccurate parameters can be obtained in a correct structure of the estimation model along with validity measures that suggest low confidence; the likely action based on the low validity measure would be to discard the correct structure in the estimation model.

These laboratory experiments indicate that statistically-derived parameters are likely to be further in error than the estimates made for system dynamics models from direct observation of processes in the actual social system. Unlike econometric models, system dynamics models are structured to facilitate parameterization by direct observation of the real-life decision-making processes. In a system dynamics model every parameter can have an independent real-life meaning. It can be discussed and evaluated in terms of its own real-life existence. Those familiar with the structure and policies of the particular part of an actual system have the necessary information to evaluate the reasonableness of a parameter value. And reasonable values are usually sufficient because behavior of a typical complex social structure is surprisingly insensitive to most parameter values (Forrester, 1961, pp. 57–59, 105, 118–119.)

2. DYNAMICS TO BE REPRESENTED

The new model of the economy addresses a wide range of dynamic behavior. The structure should be detailed enough to generate most of the major characteristics observed in the real economic system. Dynamic behavior can be described in terms of periodicities, modes, and time horizon. These characteristics are discussed below.

2.1. *Periodicities*

One might separate into individual models the different rapidities of response in a system. Each model would be designed to examine a particular kind of behavior. On the other hand, in a national socio-economic system the wide range of inherent periodicities may overlap and influence one another. Oscillatory modes of similar duration might pull together and entrain one another into a single dynamic behavior. Or they might remain separate but enhance one another's effect. Because so little is known about the diversity of such interactions, the present model combines a wide range of dynamic phenomena into one structure.

The model will simultaneously be able to create a wide span of time responses – from the business cycle on the short end, through intermediate interest-rate-capital-investment cycles, to the once-in-history transition from growth to equilibrium. To do so will require a structure containing the short time constants associated with inventories, backlogs, and bank balances. The model also represents the slower processes associated with accumulation of buildings and machinery, and the movement of people. At the long end of the time spectrum, the model should contain population growth, land occupancy, and resource depletion.

The business cycle of some three to seven years duration probably arises from interaction between inventories and employment. More commonly, the business cycle has been attributed to capital investment, but the planning time and life of capital are long enough to suggest that the primary contribution of capital investment is to a longer fluctuation in the economy. By reaching down to the fine structure of employment, inventory management, and materials procurement, the model should deal correctly with the business cycle.

In the intermediate range of behavior, some evidence suggests a cycle (the Kondratyev wave) in the economy of some fifty years duration. The existence of such a long wave is controversial. However, the structures necessary to produce such a long disturbance are to be found in labor mobility and in the processes of capital accumulation. The existence of the long wave is important to explore. If it exists, its last collapse was into the depression of the 1930's, and its next could be imminent. The long wave may have a more powerful effect than the business cycle.

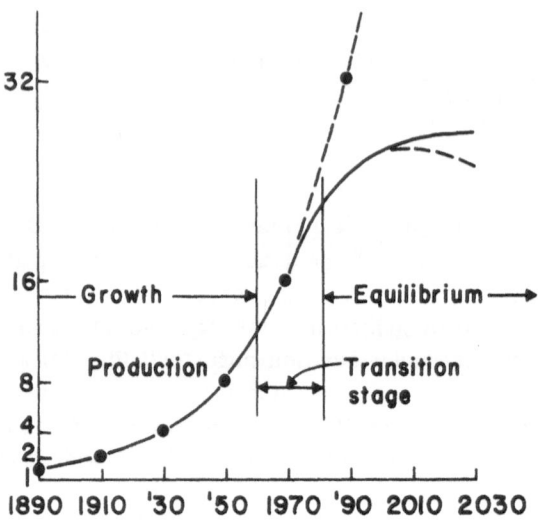

Fig. 2.   Life cycle of economic growth.

At the slowest end of the time spectrum is the life cycle of economic development. For the first time, the industrialized societies seem to be moving into a new phase of the life cycle of growth. If so, the United States is now at a point of departure from past trends and expectations such as occurs only once in the history of each civilization. As shown in Figure 2, the economy now appears to be in the 'transition stage' between growth and some form of future equilibrium in population and industrial activity. In the past, we have experienced exponential growth, with a doubling of economic output every twenty or thirty years. Such growth cannot go on forever. There must be a leveling out or a peaking and decline. The debates are about when and how the past kind of growth will end, rather than about whether or not it eventually must end.

The transition period is about half-way up the life-cycle curve of economic growth and occurs some two doubling times before the economy reaches a peak. Doubling times have been some twenty or thirty years long. At present, about fifty years before we can expect a peaking of population and economic activity, is the time when social and economic forces are pushing us out of the old growth mode.

I believe we are in the transition stage. The transition stage is consistent with the social, environmental, and inflationary forces that are developing. A model that is to cope with today's issues must incorporate, not only the recurring processes of past business-cycles, but also the transition process triggered as an economy and its population begin to fully exploit the available energy, resources, agricultural lands, and water.

The time of greatest social and economic stress occurs during the transition stage, not during equilibrium. When equilibrium has been reached, the nature of the new mode of social and economic behavior will be understood and accepted. But in the transition stage, sufficiently great forces arise to overcome the old engines of growth. Laws, attitudes, management methods, traditions, values, expectations, and religions must all change. The transition stage is the time of turbulence as the system moves out of the growth mode.

A socio-economic model to deal with today's questions should encompass the short-term dynamics of the business cycle in concert with the structures that may produce a 50-year long wave and against the background of the life cycle of economic development. For the first time we may face the triple coincidence of a business down turn, a long-wave collapse, and the pressures of the transition region. The three could combine to depress economic activity and the standard of living.

2.2. *Modes*

System modes can be described in terms of the associated restraints. The typical economic system is probably unstable in a free mid-range region where restraints are not dominating behavior. In such a restraint-free region there will be a strong tendency for the economic system to move toward one of the possible restraints. If this is correct, shifting modes of economic behavior can be described in terms of the successive restraints that dominate behavior.

At the beginning, of industrialization when there was ample land and an excess of labor, the restraint was insufficient capital equipment. The whole philosophy of capitalism arose from the capital shortage that determined the pace of economic development. But, after World War II the United States economy moved out of the mode dominated by a shortage of capital.

In the 1950's and 1960's the United States economic system has been characterized more by a shortage of manpower than by a shortage of capital. This has been indicated by chronic shortage of labor and by use of much of the capital plant only 8 hours a day instead of 24. In many businesses capital has been in sufficient supply that it is scheduled for the convenience of people. Because labor had become the principal restraint, labor demands set the style and pace of the economic system.

Now the United States is moving once again into a new mode characterized

by shifting restraints. The mode of labor shortage is giving way to the mode
of environmental shortage. Environmental shortage exists in terms of space,
agricultural land, pollution-dissipation capacity, resources, and energy.

The model should move through these various modes; not in response
to exogenous driving assumptions, but under the progression of its own
internally generated social and economic forces. In effect, the model
embodies a theory of economic development that can be tested by seeing if
it will generate the modes of behavior that have been observed.

If the model is successful in making the transitions between social and
economic modes, it should help reveal the nature of the transitions, their
causes, the forces to be expected during any specific change, and the policies
that would establish and sustain a desired mode.

2.3. *Time Horizon*

To generate the wide range of periodicities and modes, the model must be
conceived in terms of a long time horizon. The model should create appropriate
growth and fluctuation from the year 1800 to 2100. The model should
generate behavior typical of the past as a base from which to anticipate
the future. Generally speaking, forces and structures visible at any point in
time dominate a very long way into the future. If the fundamental nature of
the present system is carefully examined, most essential dynamic mechanisms
for the next several decades should be detectable.

A model with a 300-year time horizon imposes special demands on its
parameters, variables, and structure. Any constant in the model must be
constant for 300 years and must transcend the entire life cycle from early
industrial growth through all the traumatic changes of the present and near
future and into a mode of equilibrium behavior very different from the past.
Therefore, parameters must be extremely fundamental. They become descriptions of human psychology at a stable level that does not shift in response
to the immediate political and economic conditions. Any social attitudes,
ethical principles, or human preferences that themselves evolve from the
surrounding economic and geographic circumstances must be cast as variables
responsive to the socio-economic pressures. The model must be anchored on
concepts so fundamental that they represent human psychology and descriptions of nature that are not subject to change by the forces the model itself
is exploring. Concepts that are more fleeting must be formulated as variables
and generated by the forces from which they arise.

3. SOCIAL AND ECONOMIC ISSUES

A model should be constructed for specific purposes. The purposes come first and shape the design of the model. The purposes for this socio-economic model can be described in terms of the issues to be explored.

3.1. *Inflation*

The model is planned for examining the forces underlying inflation. Today's inflation is a much deeper issue than revealed by the public press or by explanations in the economics literature. It is much more than a question of inflation versus unemployment.

Present inflation arises from major imbalances in the economy. Some two-thirds of employment is outside of agriculture and direct production. This constitutes a very high overhead in government, education, and the service industry. Two-thirds of the working population in overhead is probably too great for the economy as we move out of uninhibited economic growth into a period when production is progressively more limited by environmental restraint while, at the same time, population continues to rise. Much of the inflationary pressure comes from governmental efforts to sustain a rising standard of living when real output per capita is running into inherent barriers.

The efforts to hide, by monetary and fiscal means, the fundamental changes now occuring in the industrialized economies are driving inflation. Changing social attitudes, greater complexity arising from crowding, and increasing capital investment required as space and resources become over-committed are all interlocked in the inflation syndrome.

Actions taken to counter inflation, like escalator clauses in contracts and 'indexing' of future payments to compensate for inflation, may accelerate inflation. Such changes in the legal structure of the economy should be studied in a realistic model before being put into practice.

3.2. *Recession, Depression, and Unemployment*

Most important in the near future, the model should replicate changes now occurring in the economy. By duplicating the gathering economic stresses, the model should be an effective vehicle for better understanding the causes and cures for recession coupled with inflation and a vulnerable credit and banking system.

Current national economic actions are those that might be appropriate to a normal business-cycle recession. But the forthcoming changes may be far more than a recession. If we are at the peak and entering the steep decline of a 50-year capital-investment cycle, the cause is over-investment in office buildings, automobiles, and many kinds of production facilities. Under such circumstances, national policy to sustain investment may simply delay a needed realignment within the economy.

To the extent that the final slowing of long-term economic growth is behind the present economic stresses, the fundamental issues are more demographic and environmental in nature than economic, and solutions call for redirecting the national focus of attention. A substantial percentage of the work force is devoted to creating growth and to coping with the strains arising from growth. When growth slows in the late part of the economic life cycle, substantial unemployment will develop from jobs that need no longer be filled. The day of reckoning can be delayed but not escaped. The further ahead we recognize changes imposed by the life-cycle of economic growth and work toward an orderly evolution of the socio-economic system, the less traumatic will be the realignments. Because of the extreme complexity of interactions between social, financial, technological, and demographic forces, only a comprehensive model will give access to the behavior we need to understand.

The dynamics of the Great Depression of the 1930's should be examined in search of a better understanding of causes. Was it merely from governmental mismanagement of the financial system? Was it random bad luck? Was World War I significant? Was it the collapse phase of a 50-year cycle? Did it arise as a consequence of the major migration from farm to factory? Can it recur?

3.3. *Wage and Price Controls*

The model contains separate price and wage generation in each production sector. By suppressing changes in one or more of these, the effect of price and wage controls can be examined. Prices and delivery delays (availability) influence flows in the model, so the model should realistically respond to controls and should show how controls might transform price pressures into other social and economic pressures.

3.4. *Nature of Economic Growth*

The model, by linking population, environment, knowledge generation, and technological contribution to productivity, should provide insights into the nature and future of economic growth. The world has been pursuing economic growth with success in some countries and lack of success in others. The model can be used to examine reasons for past growth and to examine whether or not the gains of the past can be sustained.

Economic growth is inherently a transient process. It cannot continue forever. But where does it lead? Will the higher standard of living be sustained in the future or fall back? The answer depends on how population, technology, and nature interrelate. The standard of living rises when production grows faster than population. But as limitations of energy, resources and space slow the growth of production, growth in population may be slowed more or less quickly with a consequent retention or loss of the standard of living achieved by past economic growth.

The end of economic growth in the equilibrium stage of Figure 2 can take many forms. If population rises faster than production in the late stages of growth, standard of living peaks in the transition stage and then declines. The end point of economic growth can move toward conditions found in India. People have striven mightily to initiate a US-type economic growth in India without success. The reason may be that India has already arrived at the end of the economic growth life-cycle – a condition in which population can exceed the capacity of the country's land and resources.

Economic growth in the United States (and, based on external resources and markets, also for Japan and Western Europe) has been a very special case. Overlooking the way the American Indian was evicted, the United States was a huge empty country of rich agricultural land and plentiful resources and energy. Economic growth has consisted of filling that land with population while using the bounty of nature. But such a growth process is a transient. It cannot continue. As growth falters, the nature of the socio-economic system changes. To what? Many choices lie in the future. Now is late but not too late to choose. But we must first have a way to examine the alternatives and the policies leading thereto.

3.5. *Agriculture*

As an economy moves through its growth life cycle, the role of agriculture changes. At the beginning the economy is rural. As capital accumulates and labor becomes scarce, and if energy is ample, agriculture becomes capital intensive. The productivity per worker in the field increases but not necessarily the output per unit of energy input. American agriculture is actually a low-efficiency converter of petroleum calories into food calories, a useful process when energy is plentiful but less effective when energy shortages develop. Toward the end of economic growth as labor becomes excessive and energy and land become scarce, a transfer of labor back to agriculture is probably necessary. Such a reversal of labor migration should be examined because it affects government policies on housing, transportation, welfare, and unemployment compensation.

3.6. *Population and Standard of Living*

As the capacity of a country becomes fully committed, a tradeoff must exist between population and the standard of living. The higher the population the lower will be the standard of living. The compromise faces each country. Whether or not the issue is recognized will mold the future character of the society. Population versus standard of living does not concern underdeveloped countries alone. Population density of Massachusetts is one and a half times the population density of India. Internal capacity of Massachusetts to support that population is probably little better than India's. Industrialized countries have been buying low-cost resources and selling high-priced manufactured exports. But as the balance shifts and resources become scarce while the capability to manufacture spreads, the status of the have and have not nations begins to converge. As every aspect of the world's capacity becomes more fully committed, each country will begin to face life within the scope of its own land and resources. Major internal realignments will be occurring. The economic mode of the future can be substantially different from the past. A national socio-economic model, if comprehensive enough, should help anticipate the actions necessary in the readjustment period.

3.7. *Education and Economic Change*

Education is a form of capital investment. Education increases skill, production output, and human satisfaction. But much of education has been used to fill the inventory of skills; with the inventory filled, only replacement is needed. The educational system, like several other parts of the services sector, shifts from being inadequate to being over-extended as the economy passes the steepest part of its growth. Evidence of excess capacity in higher education is appearing. Governmental action to withstand developing pressures and to sustain historical trends may only lead to more drastic readjustments later. By interrelating consumption demands, productivity, technology, and balance of skills, the model should generate the rising and falling balances between sectors of the economy.

3.8. *Capital Utilization*

During growth, production is primarily dependent on capital and labor. As long as land and energy are available, the standard of living rises as the capital-to-labor ratio increases. But resources, energy, and environmental capacity are consumed. In time, growth impinges on natural restraints. When the environment is fully committed, total production is limited by the capacity of nature and the standard of living is determined by the nature-to-population ratio. Under the latter circumstances, the capital-to-labor ratio becomes irrelevant to total production, which is set by environmental limits. Instead, the capital-to-labor ratio becomes a social issue. Capital intensive production with few people working can be combined with income redistribution for supporting others. Or, labor intensive production can be chosen in response to a social decision giving each person a right to a job. Such issues need classification.

3.9. *Taxes*

The consequences of collecting taxes from different points in the economy should be examined. Congress and state legislatures endlessly debate the merits and equity of who to tax and how. What are the relative advantages of property tax, personal income tax, corporate income tax, sales tax, or value-added tax. Such questions may have only a short-term significance. In the longer run, prices and wages can readjust so that money flows to the point from which it is extracted. The structure of the economy suggests that the total tax levied may be far more important than the method. If so, types of taxation may have little leverage for inducing social change, in spite of the rhetoric addressed to taxation issues. A comprehensive model incorporating various channels of taxation and containing the processes of price and wage setting should permit evaluation of tax policies.

3.10. *Balance of Payments*

As the prices of energy and resources rise relative to manufactured goods, the balance of payments deficit alters exchange rates and drives down the internal standard of living. Governmental policies affecting trade can have different long-term and short-term effects. Policies to alleviate immediate pressures can accentuate future problems. The trade-offs need to be evaluated in terms of the internal and external economic consequences. A dynamic economic model with an external sector from which to buy resources and sell goods should allow the study of national coupling to the international economy.

3.11. *Fiscal and Monetary Policy*

Economic and political debate has centered on policies for managing the economy, enhancing economic growth, and reducing unemployment. But have the policies been effective? The reduced amplitude of business cycles in the post-war years is often cited as evidence for effectiveness of such policies. But the suppressed business cycle may instead reflect other causes — the effect of governmental transfer payments, the labor-shortage mode of the economy, or the rising phase of a Kondratyev long wave caused by labor shifts and the dynamics of capital accumulation. Past policies may have contributed more to inflation than to reducing employment or stabilizing the economy.

Reviewing past policies is important, lest incorrect interpretations lead to misguided future action. With internal monetary and fiscal sectors and with taxation, debt management and government expenditures, the new economic model should offer a basis for resolving debates over Keynesian versus monetarist proposals for government intervention in the economy, and how each is related to growth, stability, and inflation.

4. STRUCTURE OF THE MODEL

The structure of the socio-economic model is intended to be general and to apply to any country having agriculture, consumption, manufacturing, and money. By concentrating first on the United States economy, while keeping in mind the desire for generality, the structure should be rich enough in detail to be a good representation of not only other industrial economies but also the underdeveloped and developing countries. Fitting the model to a particular country would merely require selection of suitable parameters and initial conditions. However, all present work is in terms of the United States.

4.1. *Overview*

The model will treat all major aspects of the socio-economic system as internal variables to be generated by the interplay of mutual influences within the model structure. The model will contain production sectors, labor and professional mobility between sectors, a demographic sector with births and deaths and with subdivision into age categories, commercial banking to make short-term loans and generate credit, savings institutions to accept saving and to make long-term loans, a monetary authority with its controls over money and credit, government services, government fiscal operations, consumption sectors, and a foreign sector for trade and international monetary flows.

A generalized production sector is being created with a structure comprehensive enough that it can be used, with selection of suitable parameters, for each of some fourteen or more producing sectors in the economy. Each sector will reach down in detail to some ten factors of production, ordering and inventories for each factor of production, marginal productivities for each factor, balance sheet and profit and loss statement, output inventories, delivery delay computation, production rate planning, price setting expectations, and borrowing.

The model is being formulated for the new DYNAMO III compiler, which handles arrays of equations and makes especially easy the replication of the production sector and its subparts. For example, an equation in the ordering function need be written only once with array subscripts to identify the ordering functions for each factor and sector.

When fully developed, the model will contain some 2000 level variables (referred to variously as integrations, state variables, stocks, or accumulations). This compares with 22 level variables in the *Urban Dynamics* model and 5 in *World Dynamics*. The total number of defined variables are about six times the number of level variables.

By reaching from national monetary and fiscal policy down to ordering and accounting details within an individual production sector, the model will bridge between the concepts of macro-structure and micro-structure in the economic system. We believe that the major modes of the economy arise from such a depth of structure and that highly realistic and informative behavior should emerge from such a degree of disaggregation.

4.2. *Standard Production Sector*

A standard production sector will be replicated to form a major part of the model. By choosing suitable parameter values, the standard sector can be used for consumer durable goods, consumer soft goods, capital equipment, building construction, agriculture, resources, energy, services, transportation, secondary manufacturing, knowledge generation, self-provided family services, military operations, and government service. Such generality focuses attention on the fundamental nature of production of goods and services and simplifies both construction and explanation of the model.

Within each production sector are inventories of some ten factors of production – capital, labor, professionals, knowledge for capital intensity, knowledge for capital productivity, buildings, land, transportation, and two kinds of materials. In addition, production is affected by length of work week for labor, length of work week for capital, and the content of each of the two kinds of materials in the product.

For each factor of production, an ordering function will create an order backlog for the factor in response to desired production rate, desired factor intensity, marginal productivity of the factor, price of the factor, price of the product, growth expectations, product inventory and backlog, profitability, interest rate, financial pressures, and delivery delay of the factor. In terms of dynamic behavior, the ordering function will be far more influential than the production function, yet, in the economics literature, attention has been in the reverse priority.

The structure of a standard production sector is essentially the structure of a single firm in the economy with parameters and nonlinear relationships chosen to reflect the broader distributions of responses resulting from aggregating together the many firms within a sector. As with a firm, the sector will have an accounting section that pays for each factor of production, generates accounts receivable and payable, maintains balance sheet variables, computes profitability, saves, and borrows money. The structure should generate the full range of behavior that arises from interactions between the real and the money and information variables. By carrying the model to such detail, it should communicate directly with the real system where a wealth of information is available for establishing the needed parameter values.

A sector will generate product price in accordance with conditions within the sector and between the sector and its customers. For testing price and wage controls, coefficients are available to inhibit price changes. The sector will distribute output among its customer sectors. Market clearing, or the balance between supply and demand, will be struck not by price alone but also on the basis of delivery delay reflecting availability, rationing, and allocation.

4.3. *Labor and Professional Mobility*

People in the production sectors are divided into two categories – labor and professional. For each category a mobility network defines the channels of movement between sectors in response to differentials in wages, availability, and need. A mobility network has a star shape with each point ending at a production sector and terminated in the level representing the number of people working in the sector. At the center of the star is a general unemployment pool, which is the central communication node between sectors. Between the central pool and each sector is a 'captive' unemployment level of those people who are unemployed but who still consider themselves a part of the sector. They are the people searching for better work within their sector or who are on temporary layoff but expecting to be rehired. In a rising demand for more labor, those in the captive level can be rehired quickly but longer time constants are associated with drawing people from other sectors by way of the general unemployment pool.

4.4. Demographic Sector

The demographic sector generates population in the model by controlling the flows of births, deaths, immigration, and aging. Age categories divide people into their different roles in the economy from childhood through retirement. The demographic sector divides people between the labor and professional streams in response to wages, salaries, demands of the productive sectors, capacity of the educational system, and family background. Workforce participation determines the fraction of the population working in response to historical tradition, demand for labor, and standard of living.

4.5. Household Sectors

The household sectors are replicated by economic category – labor professional, unemployed, retired, and welfare. Each household sector receives income, saves, borrows, purchases a variety of goods and services, and holds assets. Consumption demands respond to price, availability of inputs, and the marginal utilities of various goods and services at different levels of income.

4.6. Financial Sector

The financial sector is divided into three parts – commercial banking, savings institutions, and the monetary authority. The financial sector determines interest rates on savings and bonds, buys and sells bonds, makes long-term and short-term loans, and creates intangible variables like confidence in the banking system.

The commercial banking system receives deposits, buys and sells bonds, extends loans to households and businesses, and generates short-term interest rates. In doing so it manages reserves in response to demands of the monetary authority, and acts in response to discount rate, expected return on investment portfolio, demand for loans, and liquidity needs.

The savings institution receives savings, extends long-term loans to households and businesses, generates long-term interest rates, buys and sells bonds, and borrows short-term from the banking system. The savings institution balances money, bonds, deposits, and loans. It allocates loans between businesses and households, and it monitors the debt levels and borrowing capability of each business and household sector.

The monetary authority controls discount rate, open market bond transactions, and required reserve ratios. In doing so it responds to such variables as owned and borrowed reserves of the bank, demand deposits, inflation rate, unemployment, and interest rates.

5. STATUS, SCHEDULE, PROCEDURE

Phase One of the project has been almost entirely devoted to completing a preliminary formulation of the model that can become the basis for discussion and improvement. Such a preliminary model can be a powerful communication medium for eliciting inputs from experts. Because the model clearly reveals the assumptions made in the preliminary formulation, those with different or additional perceptions of the actual system can immediately identify errors and omissions. Preliminary model formulation is now nearly complete. At the present time (December 1974) assembly is under way and should be complete by February.

Phase Two from September 1974 through January 1975 includes a search for advisors and participants from whom to solicit suggestions for reformulation and improvements. The structure of the model and the nature of the DYNAMO III compiler permit very easy modification of the model. Because the model is so much better for communication purposes after a preliminary set of equations are operating, the most efficient time to use outside advice, criticism, and suggestion is after the preliminary model is usable. Individuals and organizations are now being identified with whom to interact to improve the model and to begin interpreting its implications.

Phase Three will extend through all of 1975 and possibly for a decade beyond. During Phase Three a widening circle of participants should become involved in a progression of discussions, model modifications, and publications on structure, behavior, and implications. As sufficient confidence in the model develops, the national issues to which it is addressed will be explored.

REFERENCES

Forrester, J. W.: 1961, *Industrial Dynamics*, M.I.T. Press, Cambridge, Mass.
Forrester, J. W.: 1969, *Urban Dynamics*, M.I.T. Press, Cambridge, Mass.
Forrester, J. W.: 1971a, *Principles of Systems*, Wright-Allen Press, Inc.
Forrester, J. W.: 1971b, *World Dynamics*, Wright-Allen Press, Inc., Cambridge, Mass.
Forrester, J. W.: 1975, *Collected Papers*, Wright-Allen Press, Inc., Cambridge, Mass.
Forrester, J. W.: 1976, 'Educational Implications of Responses to System Dynamics Models', *World Modeling: A Dialogue*, TIMS Studies in the Management Sciences, Vol. 2, North-Holland Publishing Co., Amsterdam, and American Elsevier Publishing Co., New York.

Forrester, J. W., Low, G. W., and Mass, N. J.: 1974, 'The Debate on *World Dynamics*: A Response to Nordhaus', *Policy Science*, Vol. 5, June.
Forrester, N. B.: 1973, *The Life Cycle of Economic Development*, Wright-Allen Press, Inc., Cambridge Mass.
Mass, N. J.: 1975, *Economic Cycles: An Analysis of Underlying Causes*, Wright-Allen Press, Inc., Cambridge, Mass.
Oltmans, W. L.: 1974, *On Growth*, G. P. Putnam's Sons, New York.
Senge, P. M.: 1974, 'An Evaluation of Generalized Least Squares Estimation', System Dynamics Working Paper D-1944-6, December 6, Sloan School of Management, Massachusetts Institute of Technology.

CHAPTER 7

PATTERN DISCOVERY IN ACTIVITY ARRAYS

GEORGE J. KLIR

1. INTRODUCTION

1.1. *Object, System*

The evolution of disciplinary specialization has been one of the major characteristics of the history of science. Each discipline focuses its interest on
certain kind of objects and some purpose of their investigation.

The *object* of investigation can loosely be defined as a part of the world
identified as a single entity and desirable for a particular investigation. A
hospital, an airport, a computer, a biological cell, a power station, a car, the
Sun, the Moon, a social group, a musical composition, a school, French, a
human being, etc., are examples of possible objects of investigation.

The *purpose* of investigation can be viewed as a set of questions regarding
the object which the investigator (or his client) wants to answer (Ashby, 1970).
For example, if the object of investigation is a hospital, the purpose of
investigation might be represented by questions such as "How can the ability
to give immediate care to all emergency cases be increased?", "How can the
average time spent by a patient in the hospital be reduced?", "What can be
done to reduce the cost while preserving the quality of services?"; if the investigated object is a car, the purpose of investigation might be to answer
questions "How can gas consumption be reduced?", "How can reliability be
increased?", "How can pollution produced by the car be reduced?", and the
like; if the investigated object is a musical composer, say Igor Stravinsky, the
question might be "What are the basic characteristics of Stravinsky's compositions which distinguish him from other composers?"

A basic guideline provided by systems methodology consists of a simple
but important fact that, due to restricted resources and time, objects can
almost never be studied in all their complexities. The first task in most
investigations is, therefore, to make a selection of a fairly small number of
attributes on the object which, in the opinion of the investigators, are most
relevant to the purpose of investigation.

W. E. Hartnett (ed.), Systems: Approaches Theories, Applications, 121–158.

The same attribute may possess a set of *appearances* (manifestations). If, for example, the attribute is relative humidity at a certain place, the set of appearances consists of all possible levels of relative humidity in the range from 0% to 100%; if the attribute is represented by the amount of estrogen hormone in cm^3 of blood, each particular amount is an appearance of this attribute. In most cases, the purpose of investigation does not involve the individual elements of the set of all possible appearances but rather some subsets of this set. In birth control studies, for instance, it is sufficient to identify whether the amount of estrogen hormone is greater or smaller than a specified critical level. In other cases, the individual appearances, whether significant or not, cannot be determined with complete certainty due to observation or measurement errors. Assume, for instance, that the relative humidity can be measured with accuracy of up to 1%.

The size of the considered subsets of appearances of an attribute represents a level of refinement used in observing or measuring the attribute; it is referred to as the *resolution level* of representing the attribute. Two types of entities characterize each resolution level: the original attributes selected on the object with their complete sets of appearances and new abstract attributes, referred to as *variables*, with their own sets of appearances, called *states* of the variables. Each variable is assigned to exactly one original attribute and its states represent specified subsets of the set of all possible appearances of the assigned attribute.

In the ideal case, the individual subsets of appearances of an attribute should be pair-wise disjoint so that the relation between appearances of the attribute and states of the associated variables is a mapping (many-to-one but not one-to-many). Although this relation (resolution level) is governed by the purpose for which the attribute is investigated, it must take care of unavoidable measurement or observation errors.

Even though an appropriate resolution level may eliminate most uncertainties due to measurement or observation errors, uncertainties associated with appearances close to the boundaries between the individual subsets (within the reach of measurement errors) cannot be excluded. These uncertainties can either be ignored by assuming the subsets are well defined (thus taking the risk that some empirical data may be put under wrong labels), or they can be admitted and expressed in terms of membership degree functions defining the individual subsets of appearances as fuzzy sets (Zadeh, 1965; Kaufmann, 1975). The latter approach is the basis for introducing the concept of *fuzzy resolution level* and *fuzzy variable* (Klir, 1975).

Since all attributes are observed within a certain time frame which itself must contain a specification of a time resolution level, it is necessary to distinguish between two kinds of time:

(1) *Basic time*, which may be represented either by the idealized continuous real time measured with respect to an idealized initial time instant, or by a suitable periodic time event.

(2) *Defined time*, characterized by a suitable resolution level. Either explicit or implicit modes of definition can be used. The explicit definition consists of a specification of points of basic time at which the attributes are supposed to be observed. Care must be taken in this mode of definition to make sure that distances between the specified points are sufficient enough to allow for errors in measuring the basic time. The implicit definition involves partitioning basic time into subsets by the following rule: The defined time changes whenever at least one of the variables changes its state. In the case of the explicit definition, the time resolution level is characterized by the given intervals of basic time between two successive defined times. In the case of the implicit definition, the time resolution level is fully dependent on the chosen attributes and the resolution levels at which they are observed; as such, it can be determined only after the observation.

We say that a *system* is defined, at the lowest epistemological level (level 0), on an object of investigation by: (i) a set of attributes, each associated with a set of appearances, (ii) a set of variables, each associated with a set of states, (iii) mappings from appearances of the attributes into states of assigned variables, (iv) basic time, (v) defined time.

At this level, the system represents a source of observable data defined on the object; no data are available as yet. When available, the data are recorded and processed in terms of the states of the variables involved and the defined time. To distinguish terminologically the set of attributes with their appearances and the basic time from the set of variables with their states and the defined time, they will be called, respectively, an *object system* and a *source system*. Hence, to define a system on an object, an object system, a source system and a homomorphism from the former to the latter must be defined. It may happen, although it is rather rare, that the object system and source system are isomorphic.

If we disregard the meaning of the variables and their states as well as the defined time of the source system, we say that we define a *general system* for the given source system. These two systems are isomorphic. States of variables and defined times of the general system are conveniently represented by standard sets of symbols, e.g., non-negative integers.

The described procedure of defining system on an object and the terminology introduced in the previous paragraphs are summarized in Figure 1. A more formal presentation of this procedure is included in a previous paper (Klir, 1975).

While the object and source systems are important in the process of

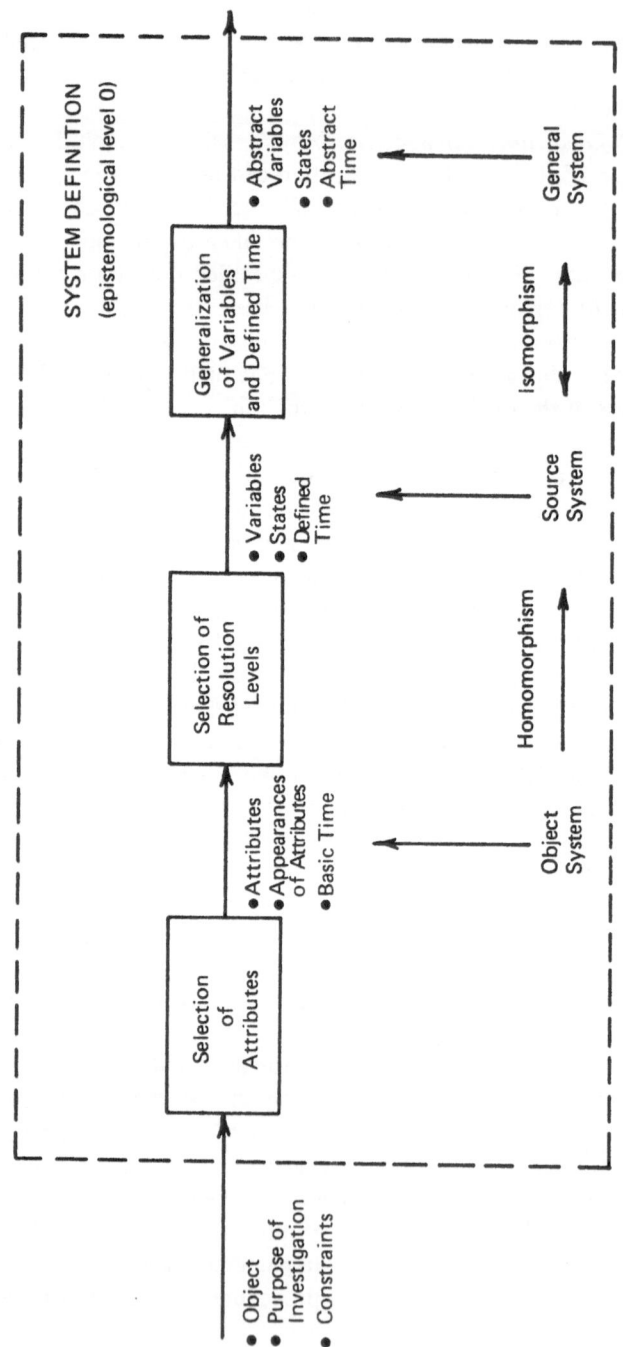

Fig. 1. The procedure of defining system on an object.

developing an operational framework for the purpose of investigation and data gathering, the general system is sufficient for any kind of data processing; the former are again needed for the interpretation of the results obtained by the data processing.

Clearly, the data corresponding to the source system can be processed essentially the same way as the data based on the general system. Due to the isomorphism between the source and general systems, no information is lost or gained when one of them is used rather than the other. The use of the general system is preferred only for its convenience; it makes the data processing standardized.

1.2. *Systems Epistemological Levels*

In the previous section, a system was defined at the lowest epistemological level (level 0) as a source of observable data. When data are available for a source system (or the general system isomorphic with it), we consider the system defined at epistemological level 1. It will be shown that the data can be conveniently organized in the form of two-dimensional arrays (matrices) or three-dimensional arrays; they are referred to as *activity arrays*.

Higher epistemological levels involve knowledge of some time-invariant relational properties representing the data. At level 2, a behavior or a state-transition relation are involved.

The behavior is defined, generally, as a relation among some present, and/or past, and/or future states of the variables which is time-invariant within the available data, and for the chosen pattern of samples taken from the data. The pattern of samples, which is referred to as the *mask*, specifies those past, present or future states of the variables which are required to be a part of each sample. As such, the mask represents a hypothetical memory structure of the system.

The *state-transition relation* (ST-relation, in abbreviation) is defined as a binary relation representing pairs of successive samples which is time-invariant within the available data and for a chosen mask. Each pair of successive samples represents a possible transition from one memory state to another.

At level 3, the system consists of a set of systems defined at levels 2, 1 or 0, referred to as elements of the larger system, and a set of couplings between them. A coupling between two elements is represented by variables they share.

At level 4, the system consists of a set of systems defined at any of the lower levels and a time-invariant procedure which describes transitions between the lower level systems.

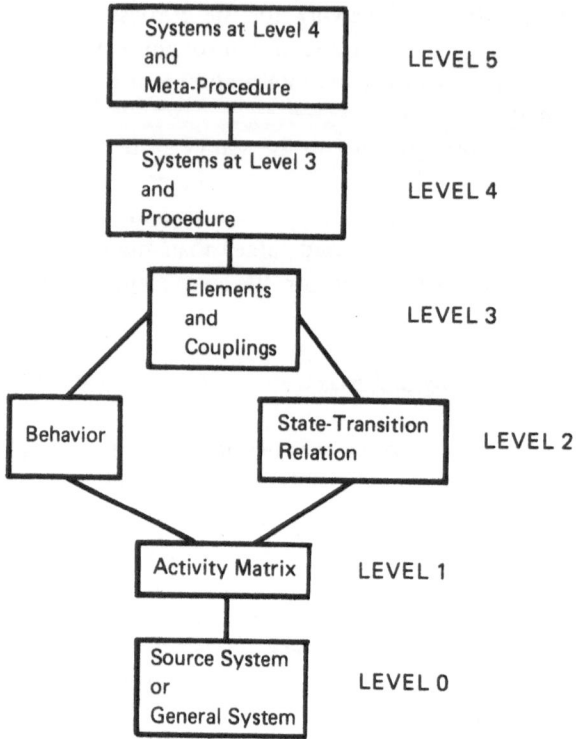

Fig. 2. Hierarchy of systems epistemological levels.

Higher epistemological levels are defined recursively by assuming a set of systems at level 4 and a time-invariant meta-procedure describing transitions between the procedures of systems at level 4 (this forms a system defined at level 5), a set of systems at level 5 and meta-meta-procedure describing transitions between the meta-procedures of systems at level 5 (this forms a system defined at level 6), etc.

The epistemological levels are summarized in Figure 2. They are ordered in the sense that knowledge at each level entails knowledge at all lower levels but contains some additional knowledge which cannot be derived from the latter. For a more comprehensive presentation of systems levels see Klir (1969), Orchard (1972) and Zeigler (1974).

1.3. *Knowledge Acquisition*

Basic classes of systemic problems can be viewed as transitions from one epistemological level to other levels. For instance, systems synthesis is a transition from level 1 or 2 to level 3 or higher levels; systems analysis (in the narrow sense) is associated with a transition from a given level to some lower levels, e.g., from level 4 to level 3; knowledge acquisition is characterized by attempts to climb up the hierarchy of epistemological levels.

There are two basic modes of knowledge acquisition which have been referred to as the discovery approach and the postulational approach (Zeigler, 1974). Starting from the empirical data (level 1), the discovery approach is viewed as a transition to level 2 or higher levels based solely on the empirical data. The postulational approach, on the other hand, starts from a system constructed at a level higher than 1 and this system is then validated at the observational level (level 1); that is to say, the system is first postulated at an appropriate level and then checked against the data rather than having been discovered from the data.

Although the discovery and postulational approaches, in their 'pure' form, are idealizations with little practical meaning, procedures which are close to one or the other form a methodological core of knowledge acquisition. Each of these procedures has some advantages and some disadvantages as pointed out by Zeigler (1974).

It seems that both of the approaches should ultimately be integrated in one powerful methodological package for knowledge acquisition. To achieve such a goal, however, it is necessary to understand first the components of this package. The aim of this paper is to present some procedures which are close to the discovery approach and are applicable to a broad class of systems. The paper treats only the transitions from level 1 to level 2; transitions to higher level are subject of further research.

1.4. *General Procedure of Discovery Approach*

The definition of a system on the object of investigation, at the epistemological level 0 (Figure 1), is the first step in the procedure of empirical investigation, as shown in Figure 3. It is followed by data gathering whose result is an activity array (epistemological level 1).

When the discovery approach is used, the next step is a processing of the activity array; its purpose is to discover suitable patterns in the activity array through which the latter can be generated. In other words, the purpose of data processing is to identify some time invariant relational properties of the data, in the form of a behavior or ST-relation, which may serve as a generator of the data. This accomplishes a transition to the epistemological level 2. The time-invariant properties are then interpreted from the standpoint of the

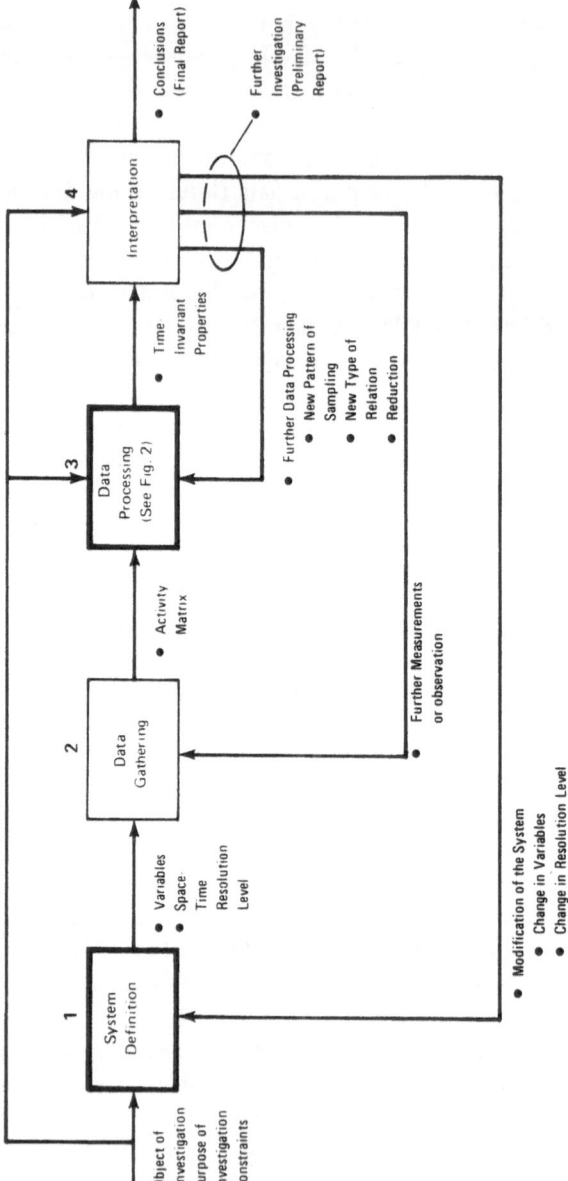

Fig. 3. The procedure of empirical investigation.

purpose of investigation and either some final conclusions are derived or further investigation is initiated restarting at step 1, 2 or 3 (Figure 3).

A procedure for processing activity arrays, which is the essence of the transition from epistemological level 1 to level 2 in the discovery approach, involves the following steps:

(1) A mask is selected as a paradigm for representing the activity array.

(2) All possible samples or all possible pairs of successive samples are determined in the activity array for the chosen mask within the whole range of defined time. The total numbers of the individual samples or their successive pairs (transitions between samples) are also determined and used for the calculations of probabilities of samples or their transitions.

(3) The results obtained by the sampling procedure are evaluated. If they are satisfactory, the data processing terminates and the interpretation of the results in terms of the object and source systems takes place. If the results are too complex to provide the investigator with any insight, they can be reduced in many different ways, providing the investigator with a spectrum of simplified representations of the data, all based on the same mask. Furthermore, the whole procedure can be repeated for several different masks and the results obtained for individual masks compared by some objective or subjective criteria.

The individual steps of the procedure of processing activity arrays, as shown in Figure 4, are described in detail in subsequent sections of this article.

In developing the procedure for data processing, systems whose variables are classified into input and output variables are distinguished from systems where such classification is not given. The latter are called *neutral systems*, the former are referred to as *controlled systems*. To be able to view a system as a controlled system, we must know which variables are controlled independently of the set of investigated variables, possibly by the observer, and which are controlled through their relationship to the other investigated variables. The former are *input variables*, the latter are *output variables*. If this classification of variables cannot be clearly established, the system must be considered a neutral system.

2. SAMPLING PROCEDURE

2.1. *Neutral Systems with Well Defined Variables*

Let $v_0, v_1, ..., v_n$ be abstract variables of a general system defined on the object of investigation, and let $V_0, V_1, ..., V_n$ be, respectively, sets of states

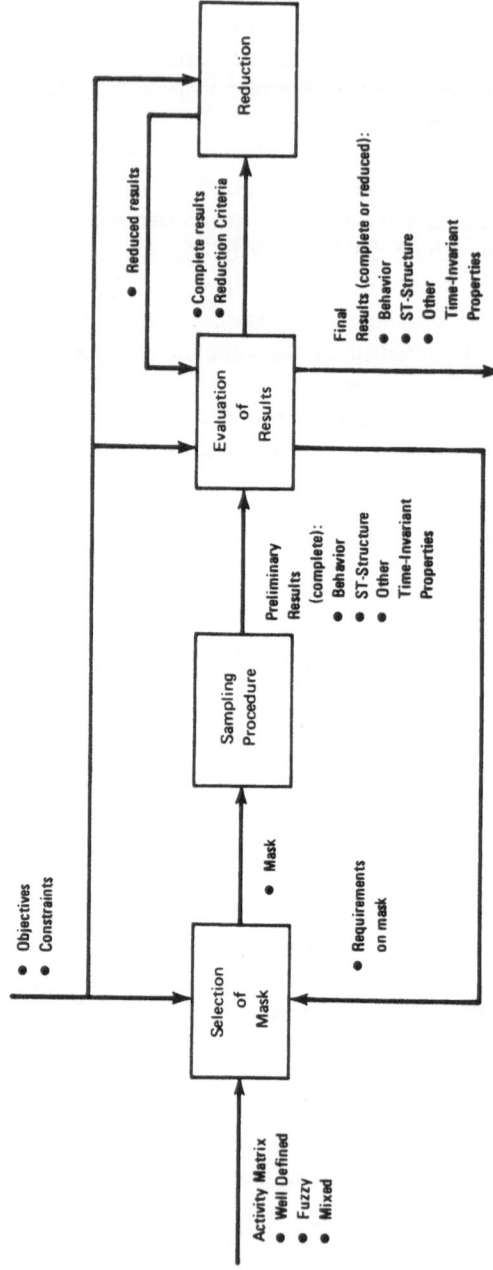

Fig. 4. The procedure of processing activity arrays in the discovery approach.

of these variables. Assume that the states are represented by non-negative integers so that $V_i = \{0, 1, ..., r_i\}$, $i = 0, 1, ..., n$, where r_i stands for the largest integer in set V_i. Let t represent defined time identifier, and T stand for the set of all defined times at which observation takes place ($t \in T$). Assume that non-negative integers are used for the defined times so that $T = \{0, 1, ..., r_t\}$, where r_t stands for the largest integer in T. Assume, further, that elements of T are ordered the same way as the corresponding intervals or instants of real time, i.e., if $t_1 < t_2$ ($t_1, t_2 \in T$), then the interval or instant of real time labelled by t_1 precedes the interval or instant labelled by t_2.

If all variables are well defined and no classification into input and output variables is meaningful, the empirical data can be viewed as a matrix $[v_{i, t}]$, where $v_{i, t}$ stands for the state of variable v_i at time t ($v_{i, t} \in V_i; i = 0, 1, ..., n$; $t = 0, 1, ..., r_t$). This $n + 1$ by $r_t + 1$ matrix (or two-dimensional array) will be referred to as the *activity matrix*.

The empirical data, represented by an activity matrix, can be processed in many different ways, each based on certain patterns of samples taken from the activity matrix. A pattern of sampling is defined by a selected set of variables s_k ($k = 0, 1, ..., q$), referred to as *sampling variables*, which are defined by the equation

$$s_{k, t} = v_{i, t+a}, \tag{1}$$

where a is an integer and $s_{k, t}$ stands for the state of sampling variable s_k at time t. A set of pairs (i, a) through which the identifiers k of sampling variables are uniquely defined is referred to as the *mask* (Klir, 1969, Orchard, 1972). It represents a two-dimensional pattern which is used for taking samples from the activity matrix. Let $s_{k, t} \in S_k$; then, clearly, $S_k = V_i$ if k is identified by a pair (i, a), for any a, through Equation (1).

A mask can be conveniently represented by a matrix in which the rows and columns are labelled by the i's and a's, respectively, and the entries are identifiers k or blanks (Figure 5(a)). The column labelled $a = 0$ in the mask is called the *reference*. When the mask is applied to an activity matrix, it produces a sample $c \in C = S_0 \times S_1 \times ... \times S_q$. The location of the reference of the mask identifies a particular time t_c associated with the sample c. For instance, the location of the mask on the activity matrix as shown in Figure 5(b) produces sample $c = (2, 0, 1, 2, 0, 1, 0, 3)$ associated with time $t_c = 5$. Detailed meanings of individual components of c in terms of Equation (1) are listed in Figure 5(c).

An exhaustive temporal sampling in the activity matrix for a given mask, through which a set of sampling variables $s_0, s_1, ..., s_q$ is defined, yields time-invariant relations

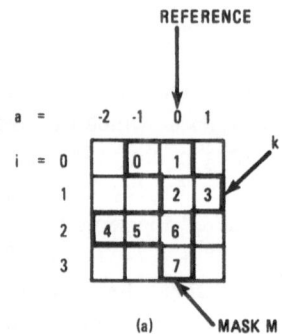

(a)

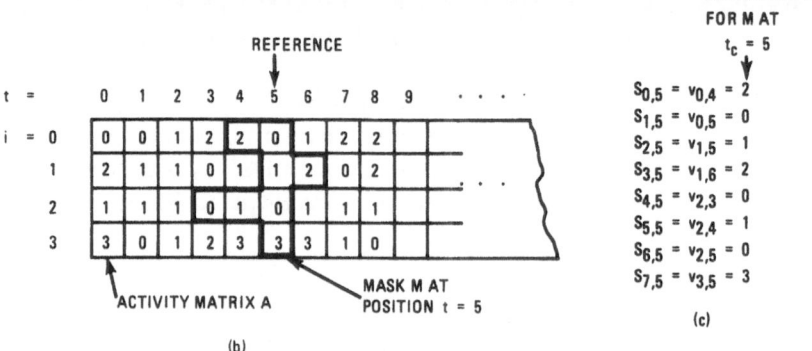

(b)

SAMPLE c OF A
FOR M AT
$t_c = 5$

$s_{0,5} = v_{0,4} = 2$
$s_{1,5} = v_{0,5} = 0$
$s_{2,5} = v_{1,5} = 1$
$s_{3,5} = v_{1,6} = 2$
$s_{4,5} = v_{2,3} = 0$
$s_{5,5} = v_{2,4} = 1$
$s_{6,5} = v_{2,5} = 0$
$s_{7,5} = v_{3,5} = 3$

(c)

Fig. 5.   Example of mask and activity matrix.

$$R_1 \subset S_0 \times S_1 \times \ldots \times S_q = C$$

and

$$R_2 \subset C \times C.$$

Elements of R_2 are pairs (c, c') of successive samples which exist in the activity matrix. Probabilities of elements of either of these relations can be calculated using standard procedures.

Let sample c appear in the activity matrix at $N(c)$ different time locations (different meaningful positions of the mask on the activity matrix). Then probability $P(c)$ with which this sample appears in the activity matrix is given by the formula

$$P(c) = \frac{N(c)}{\sum_b N(b)}. \tag{2}$$

Clearly,

$$\sum_c P(c) = 1. \tag{3}$$

The set

$$B = \{(c, P(c)) \mid c \in R_1, \ P(c) \text{ given by (2)}\} \tag{4}$$

is called a *basic behavior* of the system during the period of observation with respect to (or from the viewpoint of) the selected mask through which samples c are defined.

Let a pair of successive samples (c, c') appear in the activity matrix at $N(c, c')$ different time locations. Then probability $P(c, c')$ with which this pair of successive samples appears in the activity matrix is given by the formula

$$P(c, c') = \frac{N(c, c')}{\sum_{b, b'} N(b, b')}. \tag{5}$$

Clearly,

$$\sum_{c, c'} P(c, c') = 1 \tag{6}$$

and

$$\sum_b P(c, b) = P(c) \tag{7}$$

$$\sum_b P(b, c') = P(c'). \tag{8}$$

Also

$$\sum_b P(c, b) = \sum_b P(b, c) \tag{9}$$

except for small deviations (which are negligible if the total number of samples is large) associated with the first and last samples in the activity matrix. It can also easily be shown that

$$\sum_{c, c'} P(c) \cdot P(c') = 1. \tag{10}$$

The set

$$S = \{((c, c'), P(c, c')) \mid (c, c') \in R_2, P(c, c') \text{ given by (5)}\} \tag{11}$$

is called the *state-transition structure* (or *ST-structure*, in abbreviation) of the system during the period of observation from the viewpoint of the selected mask.

Clearly, different basic behaviors and ST-structures are obtained for different sets of sampling variables. The investigator of the system can thus select the mask so as to incorporate aspects he deems desirable. If he can define an objective function on the set of basic behaviors and/or ST-structures derivable from the activity matrix, and if he can establish some constraints regarding the selection of sampling variables, the selection of the best mask can be viewed as an optimization problem.

The ST-structure S is a powerful tool for describing complex dynamic processes. One possible representation of S is a square matrix (*ST-matrix*) in which the rows and columns are associated, respectively, with c and c', and the entries are basic probabilities $P(c, c')$ or conditional probabilities, given by the formula

$$P(c'|c) = \frac{P(c, c')}{P(c)}, \tag{12}$$

where $P(c)$ is given by Equation (7). Clearly,

$$\sum_{c'} P(c'|c) = 1. \tag{13}$$

Given a mask and an observed variable v_i, the *left-most* (or the *right-most*) *sampling variable* defined in terms of v_i by Equation (1) is that sampling variable which is assigned to the smallest (or the largest, respectively) a for the given i.

Let c_ϱ and c_r denote, respectively, those portions of c which are represented by all left-most and all right-most sampling variables associated with a particular mask. Let c_0 denote that portion of c which is represented by all sampling variables except the left-most and right-most. Then probabilities $P(c) = P(c_\varrho, c_0, c_r)$, involved in the behavior, can be easily modified to probabilities $P(c_r|c_\varrho, c_0)$ by the formula

$$P(c_r|c_\varrho, c_0) = \frac{P(c_\varrho, c_0, c_r)}{\sum_b P(c_\varrho, c_0, b)}. \tag{14}$$

The set

$$G = \{(((c_\varrho, c_0), c_r), P(c_r|c_\varrho, c_0))|(c_\varrho c_0 c_r) = \\ c \in R_1, P(c_r|c_\varrho, c_0) \text{ given by (14)}\} \tag{15}$$

is a modified form of behavior. It will be referred to as the *generating behavior* because it represents a rule of generating activity matrices based, for the given mask, on the same properties as the empirical activity matrix. Clearly,

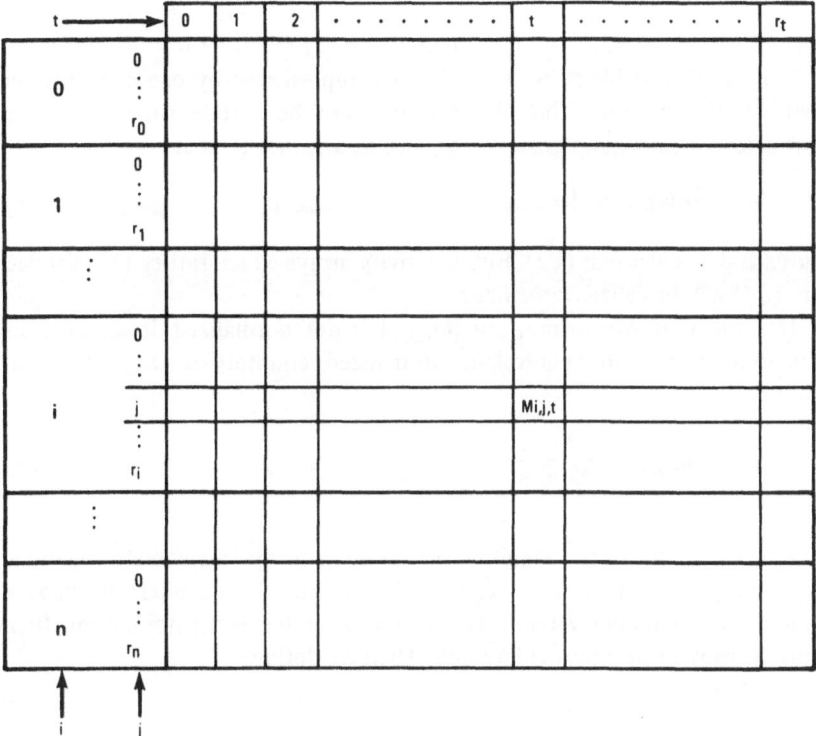

Fig. 6.   Fuzzy activity array.

$$\sum_{c_r} P(c_r \mid c_\varrho, c_0) = 1 \tag{16}$$

for each particular (c_ϱ, c_0). One possible representation of G is a matrix (*G-matrix*) whose rows and columns are associated, respectively, with (c_ϱ, c_0) and c_r. In contrast with the ST-matrix, the G-matrix is usually not a square matrix.

2.2. *Neutral Systems with Fuzzy Variables*

Let us now consider the case in which the observed variables, as introduced in Section 2.1, are fuzzy. The empirical data can be viewed as a *three-dimensional activity array* $[m_{i,j,t}]$, where $m_{i,j,t}$ stands for the membership grade of state j of variable v_i at time t $(0 \leqslant m_{i,j,t} \leqslant 1; i = 0, 1, \ldots, n; j = 0, 1, \ldots, r_i; t = 0, 1, \ldots, r_t)$. This array is formed by $n + 1$ matrices (pages), each identified by a particular identifier i; page i has $r_i + 1$ rows, identified by index j, and $r_t + 1$ columns, identified by index t (Figure 6).

Since each variable v_i at every time t is represented by one and only one state $j \in V_i$, although the observer may not be certain which one it is, normalized membership grades $m_{i,j,t}$ will be assumed so that

$$\sum_j m_{i,j,t} = 1 \tag{17}$$

for each particular pair (i, t). Fuzzy activity arrays which satisfy (17) for each pair (i, t) will be called *normalized*.

If a fuzzy activity array, say $[m'_{i,j,t}]$ is not normalized, it can be easily transformed into the equivalent normalized counterpart $[m_{i,j,t}]$ by the formula

$$m_{i,j,t} = \frac{m'_{i,j,t}}{\sum\limits_j m'_{i,j,t}} . \tag{18}$$

Let $m_{c,t}$ and $m_{c_k,t}$ stand, respectively, for the membership grade of $(q+1)$-tuple $c = (c_0, c_1, ..., c_q)$ at time t and the membership grade of state c_k of sampling variable s_k at time t in the given normalized fuzzy activity array using a particular mask. Then we define

$$m_{c,t} = \prod_k m_{c_k,t}. \tag{19}$$

For each time t, the membership grade of c is calculated by taking the product of membership grades for $c_0, c_1, ..., c_q$. This is justified by the following properties of Equation (19):

1. $m_{c,t} = 0$ iff $m_{c_k,t} = 0$ for at least one k.

2. $m_{c,t} = 1$ iff $m_{c_k,t} = 1$ for all k.

3. $m_{c,t}$ is equally proportional to each $m_{c_k,t}$ $(k = 0,1, ..., q)$.

4. If all components of c are fixed except one, say c_d, then $m_{c,t}$ is linearly proportional to $m_{c_d,t}$, i.e.,

$$m_{c,t} = b \cdot m_{c_d,t},$$

where

$$b = \prod_{k \neq d} m_{k,t}$$

is a constant.

5. $m_{c,t}$ is normalized, i.e.,

$$\sum_c m_{c,t} = 1. \tag{20}$$

This property follows directly from the facts that when combining Equations (19) and (20), Equation (20) can be written as

$$\sum_c m_{c,t} = \prod_k \sum_j m_{c_k=j,t},$$

and that membership grades of all variables are normalized (Equation (17)) so that

$$\sum_j m_{c_k=j,t} = 1$$

for each $k = 0, 1, ..., q$.

Probabilities of individual samples c can be calculated by

$$P(c) = \frac{\sum_t m_{c,t}}{N}, \tag{21}$$

where N is the total number of all meaningful positions of the mask on the activity array. Obviously,

$$N \leqslant r_t + 1,$$

the equality being meaningful if either $a = 0$ for all elements of the mask, or appropriate initial and terminal conditions of the activity array are known; e.g., the array is periodic and one period is given.

Let c' stand for a sample succeeding sample c, and let $M_{c,c',t}$ denote the membership grade of the transition from sample c to sample c' which takes place at time t. Then, using the same justification as before, we define

$$M_{c,c',t} = m_{c,t} \cdot m_{c',t+1}. \tag{22}$$

It follows from Equation (20) that

$$\sum_{c,c'} M_{c,c',t} = \sum_c m_{c,t} \cdot \sum_{c'} m_{c',t+1} = 1.$$

Probabilities $P(c, c')$ of individual transitions from c to c' can be calculated by

$$P(c, c') = \frac{\sum_t M_{c,c',t}}{N}, \tag{23}$$

where N is the total number of all meaningful pairs of successive positions of the mask on the activity array.

It is important to notice that the original uncertainties embodied in the observation process and represented by membership grades $m_{i, j, t}$ are not explicitly included in the behavior or ST-structure. However, they are implicitly incorporated in the probabilities $P(c)$ and $P(c, c')$ given, respectively, by Equation (21) and Equation (23). Rather than counting the number of appearances of sample c or a pair of successive samples (c, c'), as we do in the case of well defined variables (Equation (2) and (5)), we add membership grades associated with c or (c, c') for all times t. Equations (21) and (23) thus represent generalized definitions of probabilities for fuzzy events.

2.3. *Controlled Systems*

When the observed variables are classified into input and output variables, it is assumed that the inputs are controlled by an agent outside the object system. This agent is usually referred to as an *environment* of the system under investigation. By definition, properties of the environment – e.g., its behavior or ST-structure – are either not known or are not of interest within the context of the investigation. If they were of interest and known, they would become a part of the investigated system.

The sampling procedures for neutral systems, as described in Sections 2.1 and 2.2, can be preserved for controlled systems subject to one modification: Instead of using statements such as "the probability of transition from c to c' is $P(c, c')$", conditional statements must be used, such as "if the environment produces the state of the input variables (*stimulus, excitation*) e, then the probability of transition from c to c' is $P(c, c'|e)$." Clearly,

$$\sum_{c, c'} P(c, c'|e) = 1 \qquad\qquad (24)$$

for each particular e. Symbols c and c' stand, respectively, for the present state and the succeeding state of all sampling variables defined through the chosen mask, except the input variables; e is used for the present state of the input variables.

If probabilities conditional with respect to the present state are desirable for controlled systems, their form is $P(c'|c, e)$. They satisfy the equation

$$\sum_{c'} P(c'|c, e) = 1 \qquad\qquad (25)$$

for each particular pair (c, e).

ST-structures of controlled systems can conveniently be represented by three-dimensional arrays in which the entries are either probabilities $P(c, c'|e)$ or $P(c'|c, e)$. Let matrices of the three-dimensional array be identified by e, and let rows and columns in each matrix be identified by c and c', respectively. Then, from the computational point of view, neutral systems can be treated as a special case of controlled systems in which only one stimulus is applied which never changes. However, from the epistemological point of view, neutral and controlled systems are essentially different. In controlled systems, the environment is clearly separated from the system − i.e., we know which variables are controlled by the environment and which are controlled by the system. In neutral systems, this separation does not exist or, at least, it is not known.

For basic behaviors of controlled systems, we can use conditional statements: "If the environment produces stimulus e, then the probability of sample (state) c is $P(c|e)$". Clearly,

$$\sum_c P(c|e) = 1 \tag{26}$$

for each particular e; also,

$$P(c|e) = \sum_{c'} P(c, c'|e). \tag{27}$$

Matrices is which the rows and columns are assigned by stimuli e and states c, respectively, offer a convenient representation for basic behaviors of controlled systems.

For generating behaviors of controlled systems, the appropriate conditional statement is: "If the environment produces stimulus e and if the left-most and middle portions of the sample are c_ϱ, c_0, then the probability of the right-most portion of the sample being c_r is $P(c_r|e, c_\varrho, c_0)$". These probabilities are calculated from probabilities $P(c|e) = P(c_\varrho, c_0, c_r|e)$ by the formula

$$P(c_r|e, c_\varrho, c_0) = \frac{P(c_\varrho, c_0, c_r|e)}{\sum_b P(c_\varrho, c_0, b|e)}. \tag{28}$$

They must satisfy the equation

$$\sum_{c_r} P(c_r|e, c_\varrho, c_0) = 1. \tag{29}$$

Similar to ST-structures, generating behaviors of controlled systems can conveniently be represented by three-dimensional arrays in which the entries are probabilities $P(c_r|e, c_\varrho, c_0)$; matrices of the three-dimensional array are identified by e, rows of the matrices by (c_ϱ, c_0) and their columns by c_r.

2.4. *Behavior Versus ST-Structure*

Using Equation (7), the basic behavior of a neutral system can be uniquely determined from its ST-structure. In the case of a controlled system, the same can be done by using Equation (27). In both cases, the behavior and the ST-structure are defined for the same mask.

Furthermore, given the basic behavior of a neutral system, its generating behavior can be uniquely determined by Equation (14); in the case of controlled systems, Equation (28) must be used instead. The same mask is used for both the basic behavior and the generating behavior.

Hence, given the ST-structure of a system, both its basic and generating behaviors can be uniquely determined without any adjustment of the mask. The opposite procedure, in which the ST-structure is to be determined on the basis of either of the two kinds of behaviors is possible too but requires a modification of the mask. Some new concepts must be introduced before this procedure can be described.

Let mask M be defined as a set of ordered pairs (i, a) and let a segment of M be any sequence of ordered pairs

$$(i, \alpha), (i, \alpha + 1), \ldots, (i, \alpha + r)$$

such that

$$(i, \alpha) \in M$$
$$(i, \alpha + 1) \in M$$
$$\ldots \ldots \ldots \ldots$$
$$(i, \alpha + r) \in M$$

and

$$(i, \alpha - 1) \notin M$$
$$(i, \alpha + r + 1) \notin M.$$

Let $(i, \alpha + r + 1)$ be called the *extension* of segment $(i, \alpha), (i, \alpha + 1), \ldots,$ $(i, \alpha + r)$ and let mask ^eM, formed by all segments of mask M and all extensions of these segments, be called the *extended mask* with regard to M. Finally, let c and ^ec denote, respectively, a sample of mask M and a sample of the extended mask ^eM such that ^ec becomes c if states of sampling variables based on extensions of all segments of M are ignored. The left, middle and right portions of both c and ^ec will be distinguished, respectively, by subscripts ℓ, 0 and r.

Using the concept of the extended mask, it can be demonstrated that the ST-structure for a given activity array and mask M can be uniquely determined from the basic behavior representing the same activity array and based on the extended mask ^{e}M. Indeed,

(i) given a pair (c, c') of successive samples, we can write $(c, c') = (c_{\varrho}, c_0, c_r, c'_{\varrho}, c'_0, c'_r)$, where $(c_0, c_r) = (c'_{\varrho}, c'_0)$, and ^{e}c can be uniquely determined by taking $^{e}c_{\varrho} = c_{\varrho}$, $^{e}c_0 = (c_0, c_r)$ or $^{e}c_0 = (c'_{\varrho}, c'_0)$, $^{e}c_{\varrho} = c_r$;

(ii) given a sample ^{e}c, a pair of successive samples (c, c') can be uniquely determined by taking $c = (^{e}c_{\varrho}, ^{e}c_0)$ and $c' = (^{e}c_0, ^{e}c_r)$. Hence, there is a one-to-one correspondence between pairs of successive samples (c, c') for a mask M and samples ^{e}c for the extended mask ^{e}M. Consequently, $N(^{e}c) = N(c, c')$ and $P(^{e}c) = P(c, c')$. This suggests that only one sampling procedure is needed: a procedure for determining the set B of Equation (4). Both G (Equation (15)) and S (Equation (11)) can be determined from B and do not require, therefore, special sampling procedures. This is a great computational advantage because in the case of extremely large activity arrays, the sampling procedure might require a considerably larger amount of computer time than the suggested procedure of reorganizing the set B and recalculating the probabilities.

3. EVALUATION OF MASKS

3.1. *General Discussion*

Many different behaviors and ST-structures can be obtained for the same activity array, each one based on a particular mask. Theoretically,

$$2^{(n+1)(r_t+1)}$$

different masks can be defined on an activity array based on $(n + 1)$ variables and $(r_t + 1)$ defined times, but only a tiny fraction of them, those with a small number of sampling variables, are actually practical. Indeed, the purpose of processing an activity array is to find simple rules through which the activity array can be fully, or at least approximately, generated.

Although all behaviors and ST-structures determined by using the correct procedures are meaningful representations of the respective activity array, they may differ greatly in their significance with respect to the purpose of the investigation. A single mask or a few alternative masks are sometimes required for the purpose of the investigation. Whenever alternative masks are allowed, those which best satisfy the compromise between a small number of sampling variables and a good approximation of the activity array are desirable, independent of the purpose of the investigation. This compromise can also be described in terms of the reduction in the amount of uncertainty per sampling variable in the process of generating the activity array by the generating behavior or ST-structure, as compared with a completely random procedure of generating activity arrays for the given system. Any reduction in uncertainty obtained by using the generating behavior or ST-structure can also be viewed as an increase in the degree of determinism or as an increase in the empirical information preserved in the behavior or ST-structure.

   The *entropy function* (Shannon and Weaver, 1949; Feinstein, 1958) is used in this section as a reasonable measure of uncertainty through which the problem of objective evaluation and comparison of masks is explored. Although the uncertainty measure is applied exclusively to generating behaviors, the results can be utilized equally well for ST-structures employing the one-to-one correspondence between behaviors and ST-structures described in Section 2.4.

3.2. *Neutral Systems*

Let c_r denote that portion of sample c which is represented by all right-most sampling variables; let $\bar{c}_r$ denote the whole sample c except its right-most portion c_r. Then, given $\bar{c}_r$, the uncertainty $H_{\bar{c}_r}$ associated with the selection of c_r can be expressed in terms of the entropy function

$$H_{\bar{c}_r} = -\sum_{c_r} P(c_r|\bar{c}_r) \log_2 P(c_r|\bar{c}_r), \tag{30}$$

where $c_r \in V = V_0 \times V_1 \times \ldots \times V_n$. It is well known (Shannon, 1949; Feinstein, 1958) that

$$0 \leqslant H_{\bar{c}_r} \leqslant \log_2 |V|. \tag{31}$$

   The total uncertainty H_M associated with the generating behavior determined for mask M can be calculated as the sum of uncertainties $H_{\bar{c}_r}$ for individual states $\bar{c}_r$, weighted by the probabilities $P(\bar{c}_r)$ of these states, i.e.,

$$H_M = -\sum_{\bar{c}_r} P(\bar{c}_r) \sum_{c_r} P(c_r|\bar{c}_r) \log_2 P(c_r|\bar{c}_r). \tag{32}$$

Using Equation (31) and the requirement of

$$\sum_{\bar{c}_r} P(\bar{c}_r) = 1,$$

it becomes obvious that

$$0 \leqslant H_M \leqslant \log_2 |V| \tag{33}$$

independent of the probabilities $P(\bar{c}_r)$.

The *reduction of uncertainty* A_M associated with generating behavior for mask M can be expressed by the normalized difference between the maximum entropy (i.e., $\log_2 |V|$) and the actual entropy H_M, i.e.,

$$A_M = \frac{\log_2 |V| - H_M}{\log_2 |V|} = 1 - \frac{H_M}{\log_2 |V|}$$

$$= 1 + \frac{1}{\log_2 |V|} \sum_{\bar{c}_r} P(\bar{c}_r) \sum_{c_r} P(c_r | \bar{c}_r) \log_2 P(c_r | \bar{c}_r). \qquad (34)$$

An application of Equation (33) to Equation (34) yields

$$0 \leqslant A_M \leqslant 1. \qquad (35)$$

A *quality* Q_M *of mask* M with respect to generating behavior can be viewed as the reduction of uncertainty per sampling variable, i.e.,

$$Q_M = \frac{A_M}{q + 1} \qquad (36)$$

in case $q + 1$ sampling variables are defined through M.

Three examples illustrating how these measures can be used to compare masks are shown in Tables I–III. For the sake of simplicity, the first two examples are based on activity matrices consisting of a single well-defined two-valued variable. d is used as a mask identifier and represents the depth of the respective mask.

Table I compares four compact masks applied to a periodic variable with one period represented by the following sequence:

$$0\,0\,0\,0\,1\,0\,1\,1\,1\,1\,0\,1\,0\,0\,1\,1$$

In this case, while the masks for $d = 2, 3, 4$ are of no use ($A_2 = A_3 = A_4 = 0$), the mask for $d = 5$ is perfect because it represents a fully deterministic description of the activity matrix ($A_5 = 1$).

Table II uses the same masks as in Table I on another activity matrix. The probabilities in Table II were obtained through computer processing of the activity matrix which is too large to be given here. In this example, the mask for $d = 4$ is clearly inferior to all the others. The mask for $d = 5$ provides a deterministic description ($A_5 = 1$); from the standpoint of the reduction in the uncertainty per sampling variable, the mask for $d = 2$ is superior to all the others.

GEORGE J. KLIR

TABLE I

d	$P(c_r \mid c_r)$	$P(\bar{c}_r)$	H_d	A_d	Q_d
1	c_r = 0 1 .5 .5	Not Applicable	1	0	0
2	c_r = 0 1 0 .5 .5 1 .5 .5	.5 .5	1	0	0
3	c_r = 0 1 $\bar{c}_r$ = 0 0 .5 .5 0 1 .5 .5 1 0 .5 .5 1 1 .5 .5	.25 .25 .25 .25	1	0	0
4	c_r = 0 1 $\bar{c}_r$ = 0 0 0 .5 .5 0 0 1 .5 .5 0 1 0 .5 5 0 1 1 .5 .5 1 0 0 .5 .5 1 0 1 .5 .5 1 1 0 .5 .5 1 1 1 .5 .5	.125 .125 .125 .125 .125 .125 .125 .125	1	0	0
5	c_r = 0 1 $\bar{c}_r$ = 0 0 0 0 0 1 0 0 0 1 1 0 0 0 1 0 0 1 0 0 1 1 1 0 0 1 0 0 0 1 0 1 0 1 0 1 0 1 1 0 1 0 0 1 1 1 0 1 1 0 0 0 1 0 1 0 0 1 0 1 1 0 1 0 1 0 1 0 1 1 0 1 1 1 0 0 1 0 1 1 0 1 1 0 1 1 1 0 0 1 1 1 1 1 1 0	.0625 .0625 .0625 .0625 .0625 .0625 .0625 .0625 .0625 .0625 .0625 .0625 .0625 .0625 .0625 .0625	0	1	.2

TABLE II

d	$P(c_r|c_r)$	$P(\bar{c}_r)$	H_d	A_d	Q_d
1	c_r = 0 1 [.6	.9]	Not Applicable	.972	.028	.028
2	c_r = 0 1 $\bar{c}_r$ = 0 [1/3	2/3] → .6 1 [1	0] → .4	.6 .4	.551	.449	.2245
3	c_r = 0 1 $\bar{c}_r$ = 0 0 [0	1] 0 1 [1	0] 1 0 [.5	.5]	.2 .4 .4	.4	.6	.2
4	c_r = 0 1 $\bar{c}_r$ = 0 0 1 [1	0] 0 1 0 [.5	.5] 1 0 0 [0	1] 1 0 1 [1	0]	.2 .4 .2 .2	.4	.6	.15
5	c_r = 0 1 $\bar{c}_r$ = 0 0 1 0 [1	0] 0 1 0 0 [0	1] 0 1 0 1 [1	0] 1 0 0 1 [1	0] 1 0 1 0 [0	1]	.2 .2 .2 .2 .2	0	1	.2

Table III is an example based on an activity array consisting of a single two-valued and a single three-valued variable. In the third column of the table, the states of the sampling variables are listed in the natural order of their identifiers as defined in the second column.

3.3 *Controlled Systems*

Let $v_0, v_1, ..., v_m$ be input variables and $v_{m+1}, v_{m+2}, ..., v_n$ be output variables of the investigated system. Let $E = V_0 \times V_1 \times ... \times V_m$ be the set

TABLE III

d	Mask	P(c_r|c_r)	P($\bar{c}_r$)	H_d	A_d	Q_d
1	i = 0 [0] 1 [1]	c_r = 00 02 10 11 .2 .3 .1 .4	Not Applicable	1.846	.286	.143
2	$\bar{c}_r$ c_r i = 0 [0 2] 1 [1 3] a = -1 0	c_r = 00 02 10 11 $\bar{c}_r$ = 00 |.25|.5|0|.25| .2 02 |0|1/3|0|2/3| .3 10 |1|0|0|0| .1 11 |.125|.25|.25|.375| .4	.2 .3 .1 .4	1.34	.48	.121
3	$\bar{c}_r$ c_r i = 0 [0 2 4] 1 [1 3 5] a = -2 -1 0	c_r = 00 02 10 11 $\bar{c}_r$ = 0000 |0|1|0|0| 0002 |0|.5|0|.5| 0011 |0|0|0|1| 0202 |0|.5|0|.5| 0211 |.25|0|.25|.5| 1000 |.5|.5|0|0| 1100 |0|0|0|1| 1102 |0|0|0|1| 1110 |1|0|0|0| 1111 |0|2/3|1/3|0|	.05 .1 .05 .1 .2 .1 .05 .1 .1 .15	.738	.715	.119
4	$\bar{c}_r$ c_r i = 0 [0 2 4 6] 1 [1 3 5 7] a = -3 -2 -1 0	c_r = 00 02 10 11 $\bar{c}_r$ = 000002 |0|.5|0|.5| 000202 |0|.5|0|.5| 000211 |.25|0|.25|.5| 001111 |0|1|0|0| 020202 |0|.5|0|.5| 020211 |.25|0|.25|.5| 021100 |0|0|0|1| 021110 |1|0|0|0| 021111 |0|.5|.5|0| 100000 |0|1|0|0| 100002 |0|.5|0|.5| 110011 |0|0|0|1| 110211 |.25|0|.25|.5| 111000 |.5|.5|0|0| 111102 |0|0|0|1| 111110 |1|0|0|0|	.05 .05 .05 .05 .05 .05 .05 .05 .1 .05 .05 .05 .1 .1 .1 .05	.7	.729	.091

of all input states, $W = V_{m+1} \times V_{m+2} \times \ldots \times V_n$, and let C denote the set of states of all sampling variables defined for a mask M except for those states which are identical with input variables. Let $e \in E$, $c \in C$, $c_r \in W$, $\bar{c}_r \in C - W$.

The generating behavior of the system can be represented by a set of matrices, one for each input state e, in which the rows and columns are associated with states $\overline{c}_r$ and c_r, respectively, and the entries are probabilities $P(\overline{c}_r, c_r | e)$. These matrices can be transformed by standard formulas into matrices with probabilities $P(c_r | \overline{c}_r, e)$ and vectors with probabilities $P(\overline{c}_r | e)$.

For each e, the uncertainty $H_{M, e}$ associated with the generating behavior determined for mask M can be calculated by the formula

$$H_{M, e} = -\sum_{\overline{c}_r} P(\overline{c}_r | e) \sum_{c_r} P(c_r | \overline{c}_r, e) \log_2 P(c_r | \overline{c}_r, e). \qquad (37)$$

Clearly,

$$0 \leqslant H_{M, e} \leqslant \log_2 |W|. \qquad (38)$$

The total uncertainty H_M associated with the generating behavior for mask M can be calculated as the sum of uncertainties $H_{M, e}$ for individual input states e, weighed by the probabilities $P(e)$ determined from the investigated activity matrix, i.e.,

$$\begin{aligned} H_M &= \sum_{e} P(e) \cdot H_{M, e} \\ &= \sum_{e} P(e) \sum_{\overline{c}_r} P(\overline{c}_r | e) \sum_{c_r} P(c_r | \overline{c}_r, e) \log_2 P(c_r | \overline{c}_r, e). \qquad (39) \end{aligned}$$

Again

$$0 \leqslant H_M \leqslant \log_2 |W|. \qquad (40)$$

Using the same definition of the reduction of uncertainty A_M as in Section 3.1, we obtain

$$\begin{aligned} A_M &= 1 - \frac{H_M}{\log_2 |W|} \\ &= 1 + \frac{1}{\log_2 |W|} \sum_{e} P(e) \sum_{\overline{c}_r} P(\overline{c}_r | e) \times \\ &\qquad\qquad\qquad \sum_{c_r} P(c_r | \overline{c}_r, e) \log_2 P(c_r | \overline{c}_r, e). \qquad (41) \end{aligned}$$

The quality Q_M of mask M is obtained when the value of A_M calculated by Equation (41) is used in Equation (36).

3.4. Some Algorithms for Mask Evaluation

If, for any reason, the investigator decides to process the activity array for a particular mask, there is no incentive to evaluate and compare masks. If he decides to use several specific masks, he may or may not be interested in comparing their information content. In any event, all the masks which are subject to comparison are given in this case.

A problem arises when the investigator does not request any specific mask but is interested in finding a mask of modest size which preserves as much information from the processed activity array as possible. Some algorithms are suggested in this section for solving this problem. Two of them are based on the following assumption: Only masks consisting of a single block of d full columns, referred to as rectangular d-masks, are considered for $d = 1, 2, ..., D$, where D is specified by the investigator.

In the first algorithm, we determine the generating behavior G_D for the D-mask by sampling the activity array and, then, we derive generating behaviors G_d for all d-masks ($d = D - 1, D - 2, ..., 1$) from G_D.

To show that G_{d-1} can be derived from G_d, let $c = (\bar{c}_r, c_r)$ and $a = (\bar{a}_r, a_r)$, and let them denote, respectively, d-samples and $(d - 1)$-samples for the same activity array. Then $\bar{c}_r = (\bar{a}_r, a_r)$ and

$$P(\bar{a}_r, a_r) = P(\bar{c}_r) = \sum_{c_r} P(\bar{c}_r, c_r). \tag{42}$$

The second algorithm is based on sampling the activity array for individual d-masks in the increasing order of d, starting with 1-mask and terminating either with D-mask or with a particular d-mask ($d < D$) such that $A_d = 1$ and $A_{d-1} < 1$.

When the first algorithm is used, the activity array is sampled only once, for the D-mask. This is advantageous for activity arrays with extremely large sets of defined times, and it is necessary in all situations where the activity array is processed in real time without being stored for future use. The disadvantage of the first algorithm is that the sampling may sometimes be performed for an unnecessarily large mask; e.g., when $A_D = A_{D-1} = ... = A_{D-j} = 1$ and $A_{D-j-1} < 1$.

The advantage of the second algorithm is that the processing is stopped for the smallest d-mask which shows $A_d = 1$. Its disadvantage is that the sampling procedure must be applied repeatedly for all the usable masks. Of course, if the activity cannot be stored or if it is not feasible to store it, this algorithm is not applicable at all.

Regardless of which algorithm is used, the investigator ultimately obtains values of A_d and Q_d for the involved masks. If both $A_{d_1} \leq A_{d_2}$ and $Q_{d_1} \leq Q_{d_2}$, then the d_1-mask is clearly inferior to the d_2-mask and it may be excluded, without any loss of information or simplicity, from the list of considered masks. After all inferior masks are excluded, only valuable masks remain on the list. For instance, 5-mask is the only valuable mask for data in Table I; similarly, 2-mask and 5-mask are the only valuable masks for data in Table II. Additional criteria for reducing the list of valuable masks, based on values A_d or Q_d for the individual masks, may be employed by the investigator depending on his emphasis on A_d or Q_d.

More refined algorithms for mask evaluation can be developed which focus on individual sampling variables rather than blocks of variables. In each of these algorithms, the initial mask represents only the right-most sampling variables, i.e., those representing samples $c = c_r$. It is enlarged to represent one more sampling variable in each iteration of the algorithm. Let M be the mask developed after some iterations of the algorithm. If the algorithm does not terminate at this point, a set of masks is generated for mask M, each one representing one additional sampling variable selected from a set of specified sampling variables. Entropies of generating behaviors are calculated for each of these masks and the mask with the smallest entropy is selected as the next mask M'. Its reduction of uncertainty and quality are calculated and the algorithm either terminates or M' becomes M and the algorithm is repeated again. It terminates when either zero entropy is reached or some requirement regarding the terminal mask is satisfied, e.g., the mask represents a specified number of sampling variables, the reduction of uncertainty reaches a certain value, or the mask reaches the maximum allowed depth in the direction of time.

Two alternatives seem reasonable for specifying the set of sampling variables which can be used in extending mask M at each iteration of the algorithm:

(1) All sampling variables next to the left-most variables in mask M.

(2) All sampling variables represented by the D-mask, for a specified D, which are not represented by mask M.

The result of the algorithm is a sequence of masks, ordered in the increasing number of sampling variables, each mask M accompanied by the entropy H_M, reduction of uncertainty A_M and the measure of its quality Q_M. Although the terminal mask represents the largest reduction of uncertainty in the generating behavior (compared with the other masks in the sequence), the investigator may select another mask in the sequence instead, a mask with considerably better quality.

Although it is not guaranteed that the terminal mask in the previous algorithm represents the largest reduction of uncertainty compared with other masks of the same size selected within a given D-mask, massive experience with processing activity arrays indicates that it represents, indeed, the largest reduction of uncertainty in most cases and is close to it in the remaining cases. The chances of reaching the largest reduction of uncertainty can be improved by modifying the algorithm as follows: Given mask M developed after some iterations of the algorithm, a set of masks $M_1, M_2, \ldots$ is generated, each one representing one additional sampling variable selected from a set of specified sampling variables. Entropies of generating behaviors are calculated for all of these masks and the mask with the smallest entropy H_{min} as well as masks whose entropies H_i are close to H_{min}, say $H_i \leqslant H_{min} + \epsilon$, where ϵ is an input to the algorithm, are selected as potential masks M' for the next iteration.

Two examples of periodical activity matrices are used to illustrate the algorithms. The first example is the following activity matrix representing a single three-valued variable:

$$
\begin{aligned}
&[0\ 1\ 2\ 0\ 1\ 2\ 2\ 2\ 1\ 0\ 1\ 1\ 2\ 1\ 0\ 2\ 2\ 2\ 2\ 0\ 1\ 1\ 2\ 2\ 1\ 0\ 0\ 0\ 1\ 0\ 2\\
&\ \ 0\ 2\ 1\ 2\ 2\ 2\ 1\ 2\ 1\ 0\ 2\ 0\ 2\ 0\ 0\ 2\ 2\ 1\ 1\ 1\ 0\ 2\ 0\ 1\ 2\ 2\ 1\ 1\ 2\ 1\ 0\\
&\ \ 1\ 0\ 0\ 0\ 0\ 0\ 1\ 0\ 1\ 0\ 1\ 1\ 2\ 2\ 0\ 0\ 2\ 0\ 0\ 2\ 0\ 0\ 2\ 0\ 1\ 2\ 1\ 1\ 0\ 2\ 0\\
&\ \ 2\ 2\ 2\ 2\ 2\ 0\ 0\ 1\ 1\ 2\ 0\ 2\ 1\ 1\ 1\ 0\ 0\ 0\ 2\ 2\ 0\ 2\ 0\ 1\ 1\ 0\ 0\ 2\ 1\ 1\\
&\ \ 1\ 1\ 1\ 1\ 1\ 0\ 0\ 2\ 2\ 1\ 0\ 1\ 2\ 2\ 2\ 0\ 0\ 0\ 1\ 1\ 0\ 1\ 0\ 0\ 0\ 1\ 1\ 0\ 1\ 0\\
&\ \ 0\ 2\ 1\ 0\ 0\ 2\ 2\ 2\ 0\ 2\ 2\ 0\ 0\ 2\ 2\ 0\ 1\ 0\ 0\ 1\ 2\ 2\ 0\ 2\ 1\ 2\ 2\ 1\ 1\ 1\ 1]
\end{aligned}
$$

Let 7-mask be used as a basis within which we want to select a mask which represents the largest reduction of uncertainty for a specified number of sampling variables. The 7-mask represents 6 sampling variables $s_1, \ldots, s_6$ producing $\bar{c}_r$ and one sampling variable s_7 producing c_r; their positions within the mask are: $s_1 s_2 s_3 s_4 s_5 s_6 s_7$. The problem can be formulated as follows: Determine a mask which represents variable s_7 and g of sampling variables $s_1, \ldots, s_6$, for a particular integer $g\,(0 \leqslant g \leqslant 6)$, which is associated with the largest reduction of uncertainty in representing the given activity matrix. For all $g = 0, 1, \ldots, 6$, there are 64 different masks; they are all compared in Table IV. M is a mask identifier; rectangles in the table identify sampling variables represented by each individual mask.

Identifiers of the best mask for the individual g's are encircled in Table IV. The reader can easily verify that each of these masks can be obtained by the above described algorithm. The mask quality Q_M reaches its only maximum for $M=25$; the algorithm determines this maximum if $g \geqslant 3$ is set.

As the second example, the following activity matrix representing two three-valued variables is used:

[0 0 1 1 1 2 0 2 1 2 1 1 0 2 2 2 2 1 0 1 2 1 2 2 2 0 2 2 0
0 0 1 1 2 2 1 2 0 2 2 0 0 0 2 2 1 1 2 1 0 0 1 2 0 1 0 0 2

2 2 2 1 0 2 0 0 1 0 1 0 2 1 1 0 2 2 2 2 1 0 1 2 1 2 2 2 0
2 2 2 1 0 1 2 1 2 2 2 0 2 2 0 0 0 2 2 1 1 2 1 0 0 1 2 0 1

1 1 1 2 0 1 0 0 2 0 2 0 1 2 2 0 1 1 1 1 2 0 2 1 2 1 1 1 0
1 1 1 2 0 2 1 2 1 1 1 0 1 1 0 0 0 1 1 2 2 1 2 0 0 2 1 0 2

0 0 0 0 0 1 1 2 2 1 2 0 0 2 1 0 2 0 0 1 0 1 0 2 1 1 0 2 2
0 0 1 0 1 0 2 1 1 0 2 2 2 2 1 0 1 2 1 2 2 2 0 2 2 0 0 0 2

0 1 2 0 1 2 0 1 2 0 1 2 2 0 2 2 0 0 0 2 2 1 1 2 1 0 0 1 2
0 1 2 0 1 2 0 1 2 0 1 2 0 1 0 0 2 0 2 0 1 2 2 0 1 1 1 1 2]

Let 4-mask be used as a framework within which we want to compare the individual smaller masks. Let the position of sampling variables $s_1, s_2, \ldots, s_8$ within the mask be

TABLE IV

g	M	s_1	s_2	s_3	s_4	s_5	s_6	H_M	A_M	Q_M
0	①							1.58	.006	.006
1	2	□						1.58	.006	.003
	3		□					1.57	.012	.006
	4			□				1.58	.006	.003
	5				□			1.57	.012	.006
	6					□		1.58	.006	.003
	⑦						□	1.56	.019	.009
2	8	□	□					1.54	.03	.01
	9	□		□				1.57	.012	.006
	10	□			□			1.53	.038	.013
	11	□				□		1.54	.03	.01
	12	□					□	1.53	.038	.013
	13		□	□				1.55	.025	.008
	14		□		□			1.55	.025	.008
	15		□			□		1.53	.038	.013
	⑯		□				□	1.44	.094	.031
	17			□	□			1.53	.038	.013
	18			□		□		1.56	.019	.009
	19			□			□	1.53	.038	.013
	20				□	□		1.51	.05	.017
	21				□		□	1.5	.056	.019
	22					□	□	1.52	.044	.015
3	23	□	□	□				1.42	.107	.027
	24	□	□		□			1.34	.157	.039
	㉕	□	□			□		.429	.73	(.183)
	26	□	□				□	1.24	.22	.055
	27	□		□	□			1.42	.107	.027
	28	□		□		□		1.44	.094	.024
	29	□		□			□	1.44	.094	.024
	30	□			□	□		1.31	.176	.044
	31	□			□		□	1.4	.119	.03
	32	□				□	□	1.35	.151	.038
3	33		□	□	□			1.43	.1	.025
	34		□	□		□		1.42	.107	.027
	35		□	□			□	1.3	.182	.046
	36		□		□	□		1.35	.151	.038
	37		□		□		□	1.25	.214	.053
	38		□			□	□	1.29	.188	.047
	39			□	□	□		1.39	.126	.031
	40			□	□		□	1.31	.176	.044
	41			□		□	□	1.41	.113	.028
	42				□	□	□	1.35	.151	.038
4	43	□	□	□	□			.945	.406	.081
	44	□	□	□		□		.276	.826	.165
	45	□	□	□			□	.814	.488	.098
	46	□	□		□	□		.275	.827	.165
	47	□	□		□		□	.798	.498	.1
	㊽	□	□			□	□	.254	.84	.168
	49	□		□	□	□		.927	.417	.083
	50	□		□	□		□	1.07	.327	.065
	51	□		□		□	□	.999	.372	.074
	52	□			□	□	□	.927	.417	.083
	53		□	□	□	□		.965	.393	.079
	54		□	□	□		□	.826	.48	.096
	55		□	□		□	□	.912	.426	.085
	56		□		□	□	□	.831	.477	.095
	57			□	□	□	□	.901	.433	.087
5	58	□	□	□	□	□		.157	.901	.15
	59	□	□	□	□		□	.368	.769	.128
	(60)	□	□	□		□	□	.128	.919	.153
	61	□	□		□	□	□	.136	.914	.152
	62	□		□	□	□	□	.493	.69	.115
	63		□	□	□	□	□	.424	.733	.122
6	(64)	□	□	□	□	□	□	.06	.962	.137

$s_1 s_2 s_3 s_7$

$s_4 s_5 s_6 s_8$

All possible masks are compared in Table V, where the same notation is used as in Table IV.

We can see that the mask quality has again only one maximum for $M = 39$. We can also see that the algorithm based on adding one sampling variable in the direction of the minimal entropy in one step would determine this optimal mask. However, if we looked for the best mask with $g = 5$, the

TABLE V

g	M	s_1	s_2	s_3	s_4	s_5	s_6	H_M	A_M	Q_M
0	(1)							3.12	.016	.008
1	2	□						3.098	.023	.008
	3		□					3.083	.027	.009
	4			□				3.081	.028	.009
	5				□			3.091	.025	.008
	(6)					□		3.075	.03	.01
	7						□	3.089	.026	.009
2	8	□	□					2.94	.073	.018
	9	□		□				2.948	.07	.018
	10	□			□			2.98	.06	.015
	11	□					□	2.957	.067	.017
	12	□					□	2.958	.067	.017
	13		□	□				2.951	.069	.017
	14		□		□			2.968	.064	.016
	15		□				□	2.947	.07	.018
	16		□				□	2.947	.07	.018
	(17)				□	□		1.83	.423	.106
	18				□		□	1.88	.407	.102
	19			□			□	2.971	.063	.016
	20				□	□		1.879	.407	.102
	21				□		□	2.978	.061	.015
	22					□	□	2.941	.072	.018
3	23	□	□	□				1.678	.471	.094
	24	□	□		□			1.707	.462	.023
	25	□	□				□	1.724	.456	.091
	26	□	□				□	1.74	.451	.09
	27	□		□	□			1.607	.493	.098
	28	□		□		□		1.655	.478	.096
	29	□		□			□	1.757	.446	.089
	30	□				□	□	1.631	.485	.097
	31	□					□	1.785	.437	.087
	32	□				□	□	1.746	.449	.09
3	33	□	□	□				1.593	.497	.099
	34	□	□		□			1.657	.477	.095
	35	□	□				□	1.63	.486	.097
	36	□			□	□		1.62	.489	.098
	37	□			□		□	2.658	.162	.032
	38	□				□	□	1.69	.467	.093
	(39)				□	□	□	.459	.855	(.171)
	40			□			□	1.629	.486	.097
	41		□			□	□	1.622	.488	.098
	42			□	□	□		1.632	.485	.097
4	43	□	□	□	□			.34	.893	.149
	44	□	□	□		□		1.437	.547	.091
	45	□	□	□			□	.365	.885	.147
	46	□	□		□	□		.371	.883	.147
	47	□	□		□		□	1.428	.55	.092
	48	□	□			□	□	.398	.874	.146
	(49)	□		□	□	□		.253	.92	.153
	50	□		□	□		□	.444	.86	.143
	51	□		□		□	□	.41	.871	.145
	52	□			□	□	□	.444	.86	.143
	53		□	□	□	□		.266	.916	.153
	54		□	□	□		□	1.363	.57	.095
	55		□	□		□	□	.326	.897	.15
	56		□		□	□	□	1.38	.565	.094
	57			□	□	□	□	.291	.908	.151
5	58	□	□	□	□	□		.164	.948	.135
	59	□	□	□	□		□	.194	.939	.134
	60	□	□	□		□	□	.214	.932	.133
	61	□	□		□	□	□	.233	.926	.132
	62	□		□	□	□	□	.174	.945	.135
	(63)		□	□	□	□	□	.155	.951	.136
	(64)	□	□	□	□	□	□	.113	.964	.121

algorithm would fail – we would obtain mask 58 rather than 63. Nevertheless, the difference of quality between these two masks is very small indeed (equal to 0.001). If the modified algorithm were used with $\epsilon \geqslant 0.013$, the optimal mask would be determined for every $g = 0, 1, ..., 6$.

4. REDUCTION OF ST-STRUCTURES

4.1. Reduction Possibilities

Once the ST-structure is determined, it can be used either in its complete form or in some reduced form if this is necessary or desirable. The reduction is necessary if the number of samples is prohibitively large; it is desirable if the investigator wants to focus on certain aspects implicitly included in the ST-structure.

There are many ways to reduce ST-structures. Some are based on the investigator's subjective criteria, such as an arbitrary partitioning of states into blocks, a selection of a small subset of states, an exclusion of some sampling variables, or a reduction in the resolution level of some sampling variables. Other reductions are based on objective criteria, for instance:

(1) All states and/or transitions which appear with probability smaller than a specified threshold are excluded from the ST-structure.

(2) A specified number of states associated with the highest probabilities are left in the ST-structure; all other states are excluded.

(3) Optimal partitioning of states is determined algorithmically for a specified number of blocks in the partition, for a fixed amount of information which is allowed to get lost in the reduction process, or for similar criteria. One such algorithm, based on optimal partitioning of state sets of the individual variables, was suggested by Krippendorff (1974).

(4) States c_i, c_j are considered compatible if $|P(c_i, c') - P(c_j, c')| < K$ for all $c' \in C$, where K is a specified threshold. Maximal compatible classes (largest subsets of pair-wise compatible states) are determined and a simple minimal covering problem is solved.

The methods for reducing the ST-structure suggested above are certainly not exhaustive. Rather, they should be viewed as examples of a large variety of possibilities.

4.2. Reduction Implementation

Every approach to reducing the ST-structure leads ultimately to some combination of the following changes in the ST-structure: (i) exclusion of some

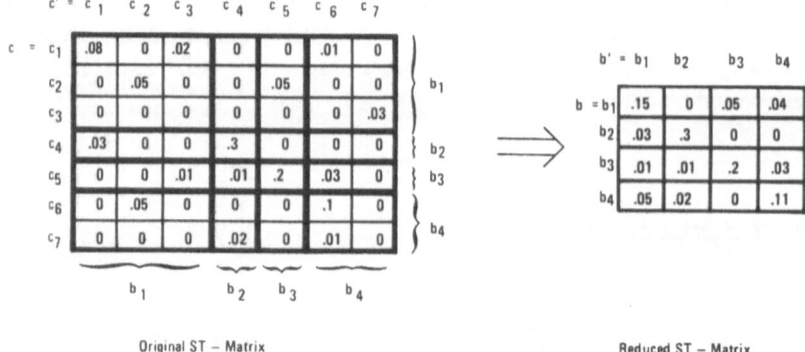

Fig. 7. Partitioning of states.

states (samples), (ii) exclusion of some transitions, (iii) partitioning of states into blocks.

If some states are excluded, the ST-matrix is reduced by eliminating rows and columns corresponding to the excluded states; the reduced matrix must then be normalized. If some transitions are excluded, the respective entries in the ST-matrix are all changed to zero and, again, the emerging matrix must then be normalized.

Let a set of states C be partitioned into set $B = \{b_1, b_2, \ldots, b_m\}$, where $m \leqslant |C|$, $b_i \subset C$, $b_i \cap b_j = \emptyset$, $\underset{i}{\cup} b_i = C$ for all $i, j = 1, 2, \ldots, m$ and $i \neq j$. Then,

$$P(b_i, b_j) = \underset{c_k \in b_i}{\Sigma} \; \underset{c_\ell \in b_j}{\Sigma} \; P(c_k, c_\ell). \qquad (43)$$

This can easily be done by partitioning the ST-matrix of basic probabilities as illustrated by an example in Figure 7.

5. CONCLUSIONS

5.1. *Basic Features of the Methodology*

A recognition of the great variety of different representations of the same empirical data is inherent in the methodology described in this article. It is reflected in the concept of mask, the possibility of representing the data by the basic behavior, generating behavior or ST-structure, and the multiplicity of possible ways to reduce behaviors or ST-structures. Deterministic, probabilistic, teleological, etc., representations are all incorporated in one complete package of structural paradigms. It is not required that the empirical data be represented by only one of the paradigms. Rather, a set of structural paradigms, each reflecting certain aspects of the data in its representation, may be used simultaneously. Together they may given the investigator much better insight than any one of them could afford alone.

If desirable, the possible representations can be compared on the basis of the amount of information regarding the empirical data preserved in the individual representations or a trade-off between the amount of information and the number of sampling variables involved. The amount of information preserved in a representation, defined in terms of the entropy function, can also be viewed as a degree of determinism in the representation. The highest degree is reached when the representation is fully deterministic; such a representation, which is not necessarily unique for the given data, provides the investigator with a perfect generator of the data.

Probabilistic representations are necessary if no deterministic representation can be obtained within the largest acceptable mask. A basis for determining a probabilistic representation with the highest possible degree of determinism, within the given constraints, is formulated in Section 3.4. A precaution must be taken to estimate errors in probabilistic representations, depending on the number of samples available in the activity arrays, on the basis of the laws of large numbers of standard probability theory (Feller, 1964). A basis for the error estimation in the entropy measures was developed by Pfaffelhuber (1971).

In addition to the recognition of the multiplicity of correct representations of the same empirical data and the possibility of determining an optimal representation within the given constraints, the suggested methodology possesses some other unorthodox features, particularly:

(1) It is applicable to variables of any scale, and to both well-defined and fuzzy variables.

(2) No *a priori* classification into input and output is required; the procedure is applicable to both neutral and controlled systems.

The described procedure is a natural one for computer implementation. While batch processing is advantageous for the sampling procedure, especially in the case of an extremely large number of defined times, an interactive mode is desirable for the reduction of behaviors or ST-structures.

It is shown in Section 2.4 that the basic behavior, generating behavior and ST-structure are, actually, different forms of the same representation. Each of them is useful at different stages of the procedure. The ST-structure is the preferred form for the investigator; it has the ability to generate the data and can be conveniently expressed in terms of stochastic matrices or ST-diagrams. The generating behavior is superior particularly for the application of entropy measures; also it has the ability of generating the data and is convenient for storing the information regarding the ST-structure. The basic behavior, although it does not possess the ability to generate the data, can easily be determined by the sampling procedure, and then transformed into the generating behavior of ST-structure.

Although the described methodology does not represent a pure discovery (or assumption-free) approach to knowledge acquisition, it is close to such approach. The only assumption used is the specification of the largest mask within which the mask optimization is performed.

5.2. *Discovery Versus Prediction*

To conclude the article, I would like to emphasize that the described procedure of processing activity or data arrays is, using Herbert Simon's (1973) terminology, a law-discovery process, i.e., "a process for recording, in a parsimonious fashion, a set of empirical data".

It is important to understand the difference between questions of law-discovery and questions of prediction. They are, unfortunately, too often mixed together. I can hardly do better than to use Herbert Simon's (1973) excellent explanation of the difference:

Law discovery means only finding pattern in the data; whether the pattern will continue to hold for new data that are observed subsequently will be decided in the course of testing the law, not discovering it. ... The discovery process runs from particular facts to general laws that are somehow induced from them; the process of testing discoveries runs from the laws to predictions of particular facts deduced from them. ... The fact that a process can extract pattern from finite data sets says nothing about the predictive power of patterns so extracted for new observations. As we move from pattern detection to prediction, we move from the theory of discovery processes to the theory of processes for testing laws. To explain why the patterns we extract from observations frequently lead to correct predictions (when they do) requires us to face again the problem of induction, and perhaps to make some hypothesis about the uniformity of nature. But that hypothesis is neither required for, nor relevant to, the theory of discovery processes. The latter theory does not assert that data are patterned. Rather, it shows how pattern is to be detected if it is here. This is not a descriptive or psychological matter, it is normative and logical. By separating the question of pattern detection from the question of prediction, we can construct a true normative theory of discovery − a logic of discovery.

In the sense argued by Simon, the methodology described in this paper should be viewed as an approach to the *normative theory of discovery*. Its extension to the questions of prediction requires further elaboration, which is beyond the scope of this article.

5.3. *Potential Extensions of the Methodology*

Although the methodology for empirical investigation, as described in this article, is restricted to the format of activity arrays and the determination of their time-invariant properties, it can be extended to other forms of data, where time is irrelevant, and other kinds of invariance, e.g., space invariance.

Let variables involved in the investigation be partitioned into variables $v_1, v_2, \ldots, v_n$, called *basic variables*, and variables $z_1, z_2, \ldots, z_m$, called *supporting variables*. While the basic variables have the same meaning as the variables in activity arrays, the supporting variables play a role similar to the defined time in activity arrays. If the variables are well defined, we can form a *data matrix*, similar to the activity matrix. Its rows and columns are assigned, respectively, to individual basic variables and states of supporting variables. If the basic variables are fuzzy we can form a three dimensional data array, similar to the activity array.

Once a data array is formed, it can be processed in exactly the same way as the corresponding activity array. Instead of time-invariant properties of all variables, we obtain properties of the basic variables invariant with respect to the supporting variables.

The major difference between the activity arrays and data arrays is that columns of activity arrays are uniquely ordered while columns of data arrays can be ordered in many different ways. A particular ordering, or a few orderings, may be required by the purpose of investigation. If it is not, we encounter a difficult problem of how to select the best ordering or a small number of good orderings.

Other extensions may include: (i) the transition, based on the discovery approach, from epistemological level 2 to level 3 or even higher levels: (ii) some aspects of prediction; (iii) integration with the postulational approach.

REFERENCES

Ashby, W. R.: 1970, 'Analysis of the System to be Modeled', in R. M. Stogdill (ed.), *The Process of Model-Building*, Ohio University Press, Columbus, Ohio.

Feinstein, A.: 1958, *Foundations of Information Theory*, McGraw-Hill, New York.

Feller, W.: 1964, *An Introduction to Probability Theory and Its Applications* (second edition), John Wiley, New York.

Kaufmann, A.: 1975, *Introduction to Fuzzy Set Theory*, Academic Press, New York.

Klir, G. J.: 1969, *An Approach to General Systems Theory*, Van Nostrand Reinhold, New York.

Klir, G. J.: 1975, 'On the Representation of Activity Arrays', *International Journal of General Systems*, 2, No. 3, pp. 149–168.

Krippendorff, K.: 1974, 'An Algorithm for Simplifying the Representation of Complex

Systems', in J. Rose (ed.), *Advances in Cybernetics and Systems*, Gordon and Breach, London.

Orchard, R. A.: 1972, 'On An Approach to General Systems Theory', in G. J. Klir (ed.), *Trends in General Systems Theory*, John Wiley, New York.

Pfaffelhuber, E.: 1971, 'Error Estimation for the Determination of Entropy and Information Rate from Relative Frequencies', *Kybernetik*, 8, No. 2, pp. 50–51.

Shannon, C. E. and Weaver, W.: 1949, *The Mathematical Theory of Communication*, The University of Illinois Press, Urbana.

Simon, H. A.: 1973, 'Does Scientific Discovery Have a Logic?' *Philosophy of Science*, December, pp. 471–480.

Zadeh, L. A.: 1965, 'Fuzzy Sets', *Information and Control*, 8, No. 3, June, pp. 338–353.

Zeigler, B. P.: 1974, 'A Conceptual Basis for Modelling and Simulation', *International Journal of General Systems*, 1, 213–228.

CHAPTER 8

A PURPOSIVE BEHAVIOR MODEL*

PETER MILNER

The title of this talk, 'A purposive behavior model', is intended to stir faint
reverberations of Tolman's *Purposive Behaviorism in Animals and Man* in
any of you old enough to have heard of it. It was published in 1932 at about
the time when I was devouring all the information I could get about robots,
and even trying to make one out of Meccano. I have heard that Tolman
started out as an electrical engineer and as I did too I have always been
sympathetic to his views. During a barbarous period in the history of psychology
he kept alive some of the progress towards an understanding of higher mental
processes that had been made by nineteenth-century associationist philosophers.
The rest of the early behaviorists were so terrified of the bogy of mentalism
that they didn't dare to even think about any behavior more complicated
than a reflex, or any part of the nervous system above the neck. Tolman
seems to have escaped the worst effects of this general phobia, perhaps
because he had a better appreciation of the capabilities of complex mechanisms.

Unfortunately Tolman had a rather obscure style, but his ideas were well
interpreted and summarized in a chapter called (confusingly for bibliographies)
'Edward C. Tolman' by MacCorquodale and Meehl (1954), and this chapter
had a profound influence on my own thinking about why we do what we do.

Various names have been given to theories that attempt to adapt association of ideas to the behavioristic form; cognitive, expectancy, or stimulus–
stimulus ($S–S$) association, to distinguish them from stimulus–response
($S–R$) association theories derived from the conditioned reflex paradigm.
The word 'stimulus', as used by both $S–S$ and $S–R$ theorists is almost always
a euphemism for idea, or concept, words that were strictly taboo to early
behaviorists. Hence $S–S$ can be directly translated as idea–idea association.

Some people criticise $S–S$ theory because it sounds as if it deals with non-
material things, but Hebb and others have shown us, in principle at least,
how objects and events can be represented by patterns of neural activity,

* Supported by grant number A66 from the National Research Council of Canada.

W. E. Hartnett (ed.), Systems: Approaches, Theories, Applications, 159–168.
Copyright © 1977 by D. Reidel Publishing Company, Dordrecht-Holland.
All Rights Reserved.

just as they can by patterns of magnetic fields, in a computer for example, and these representations correspond to concepts or ideas. There need be nothing immaterial or mystical about an idea.

A more cogent criticism of many early *S–S* theories is that they do not explain how ideas can control actions. Tolman was accused, with some justice, of leaving his rats lost in thought. MacCorquodale and Meehl addressed themselves to this problem in the chapter I mentioned, and made some additions to Tolman's theory that provide the necessary links between the conceptual and response mechanisms.

The important feature of their modification is that *response* concepts are postulated in addition to *object* concepts. In other words, an organism can not only think about food, or triangles, when no food or triangle is present, it can think about eating or walking without performing these actions. For MacCorquodale and Meehl the units of behavior are 'stimulus–response–outcome', rather than merely 'stimulus–response' as in conditioned reflex theories, or 'stimulus–outcome' as in pure *S–S* association theory.

These triads are stored by associative links in the brain so that one has to have ideas of both a stimulus and a response in order to predict an outcome. This makes sense because in a particular situation such as a road junction, or a choice point in a game, for example, several outcomes will usually have been experienced, depending on what response was made. The predicted or expected outcome is the one receiving associative input from both the stimulus situation (real or imagined) and the contemplated response.

So far we see how response ideas are needed to make predictions, but how do we get from the ideas to overt responses? This is the elegant part of the theory, permitting us to explain such things as latent learning, problem. solving, and a variety of phenomena not easily handled by classical learning theory. The fate of response ideas is determined by the value of the outcome they predict. If an outcome is noxious, the corresponding response idea is strongly inhibited, if it is boring the response idea is weakly inhibited, or at least not facilitated, but if the outcome is of value to the organism at the time, the corresponding response idea will be facilitated and raised to the status of an overt response.

I often have difficulty getting this idea across to psychologists — perhaps because too much exposure to *S–R* theory has made them unreceptive to alternatives — but it is true that some points need further explanation. In particular, the origin of response ideas needs to be considered. It is implicit in this theory that the motor systems of organism are capable of spontaneous activity (incapable of prolonged inactivity — in fact) and that all organisms bring a collection of innate and (except the newborn) experientially determined response tendencies into any new situation.

The behavior that most commonly results from these pre-existing connections is exploration or manipulation. In the case of a rat or a dog for example, it often takes the form of moving about and sniffing with the nose close to the ground, or to some object of interest. Primates more often tend to look at and handle things. The signs in shops and museums, entreating visitors not to touch, are witness to the low threshold with which this response may be elicited in man by unfamiliar objects. These useful and no doubt innate reactions are augmented by the discoveries that it is impossible to walk through solid walls, though they can sometimes be climbed or jumped, and so on. The result is that animals make a variety of overt responses in relation to any new situation, and we may assume that other response ideas spontaneously arise without becoming overt.

In a new situation, obviously none of the outcomes are known, so that if exploratory responses depended upon facilitation from desired outcomes we would never see any responding at all. This clearly does not correspond to the facts, and it suggests that the normal state of the response system is such that all response ideas become overt in the absence of inhibition. As inhibiting connections are established the animal is said to become habituated to the situation, or the exploration is said to satiate.

My theory is that concepts normally have inhibitory connections to the response system, so that as a response idea acquires conceptual associations, its likelihood of becoming an overt response diminishes. In everyday terms, when you know what you will find if you lift the lid, or turn the corner, you don't often bother to perform these operations, except in those special cases when you expect to find something you want.

Thus, control over the response system is by inhibition as well as facilitation and the primary influence of the conceptual system on responses may be via the inhibitory system. The facilitatory system may merely reflect a general condition of arousal or lethargy of the organism.

Figure 1 shows the bare essentials of the model I have been describing. The initial stimulus conditions are represented at the conceptual level by s. The various possible response ideas are labelled r_1, r_2, r_3, etc., and the outcomes o_1, o_2, o_3. We will assume that, as a result of previous experience, associations have been established between r_1 and o_1, r_2 and o_2 and r_3 and o_3, storing information about what leads to what.

To keep the diagram as simple as possible I have omitted many important features of the model; I will return to them later. As mentioned earlier, it is assumed that the o concepts have moderately strong connections to a

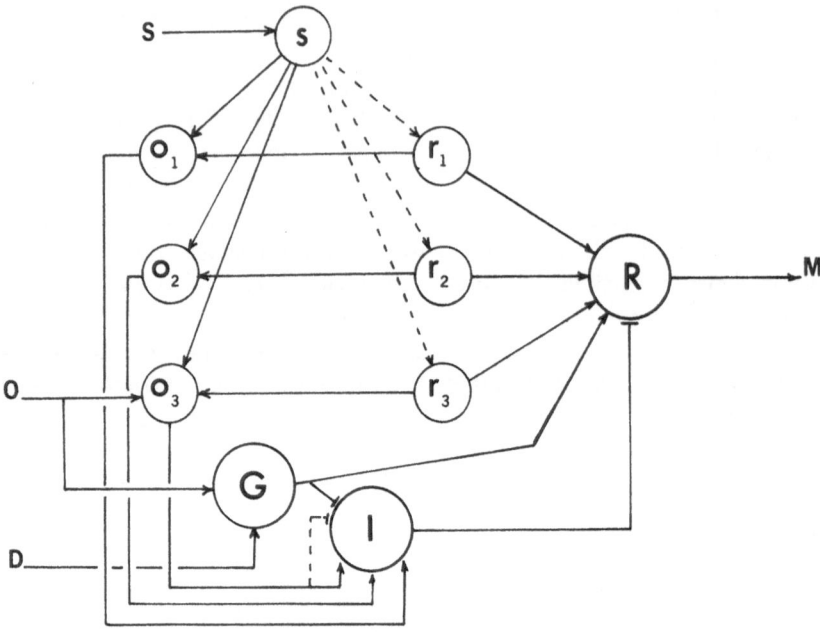

Fig. 1. A simple model based on MacCorquodale and Meehl's theory. Small circles represent concepts, *s* the concept of the incoming stimulus, *o* and *r* concepts of outcomes and responses respectively. The dotted connections from *s* to the response concepts are slowly forming associative connections. Connections between other concepts represent more rapid associations. *S*, stimulus; *O*, reward outcome; *G*, drive gate leading to motivational path; *I*, response inhibitory system; *R*, response gate leading to the motor system, *M*. The dotted inhibitory connection between the concept of the reward, o_3, and *I* is the association formed by contiguous firing of o_3 and *G*. *D* is the drive input to the drive gate.

response inhibitory system (*I*). Thus a response concept activity that fires an outcome concept initiates an inhibitory signal that prevents the response from getting to the motor system.

Let us now stipulate that one of the outcome concepts, o_3, corresponds to a reward, such as food. If the animal is food deprived, a drive signal (*D*) from the detectors of this state will open the gate *G* to food related stimuli and allow such stimuli to trigger approach and consummatory behavior. In order that the responses shall not be inhibited by concurrent concept activity, *G* also feeds an inhibitory signal to *I*. Under these conditions o_3 is excited by sensory input but is not able to fire *I*.

The next step is crucial for the performance of learned behavior. In place of the usual excitatory connection, an inhibitory association is formed between o_3 and I because o_3 fires at the time I is being inhibited by G. I have shown this new connection simply by an inhibitory addition to the previous excitatory connection, but, of course, other paths, perhaps involving interneurons might be acquired to serve this purpose. For example, o_3 might acquire an association with G as they are both fired at the same time by food stimuli.

Once the connections are formed, whenever r_3 becomes active it will fire o_3, which will have a resultant inhibitory effect on I, that is a disinhibitory effect on the response gate. Thus the disinhibitory effect of the reward will be extended to the response that led to reward, making that response more likely to occur than others.

We now have a primitive thinking machine. Placed in a situation S, it can run through a list of responses and predict the result of making each (given the necessary previous experience, of course) and select for present use the one that has proved most useful in similar circumstances in the past.

The notion of usefulness may also need some elaboration. It implies a value system and I think we must accept that some basic values are determined by innate neural connections. Forms that have survived the weeding out of evolutionary selection are those best equipped to survive and reproduce. Before individual learning evolved as an aid to survival, trial and error genetic selection was the process for improving the effectiveness of the nervous system, as well as all other physiological systems, and animals whose nervous systems did not ensure approach and eating of nourishing substances did not transmit their faulty nervous systems to any offspring. Given a number of innate survival mechanisms, other derived or secondary values can be acquired, but I do not propose to discuss this process today.

The model is quite incomplete as it stands. Avoidance of punishment requires that the normal inhibitory effect of an outcome concept be augmented if the outcome happens to be unpleasant. Passive avoidance, i.e., not making a response that is liable to hurt, can be explained in that way. Active avoidance is somewhat more complicated. It usually involves making a response that gets one out of danger, and it is usually learned on the basis of making a response that reduces fear. If fear has a combined excitatory and inhibitory effect on the response system, the excitatory effect being long-lasting, the inhibitory effect more labile, responses that lead to a safe place may become associated with a mechanism for reducing the inhibitory component and thereby acquire the ability to disinhibit themselves as rewarded responses do.

Another important process whose omission greatly detracts from the flexibility of the model as it stands is extinction, that is, the model lacks a mechanism for reprogramming itself when circumstances change. Animals do not persist indefinitely in making a response that is no longer followed by a reward. Associations once made are not immutable.

Some of the changes producing extinction apparently take place at the conceptual level (assuming that introspective evidence is to be relied upon). We know that our expectancies change with experience, usually depending upon the recency and frequency of the experience. If I make a well-established response and get an unexpected outcome my reaction is astonishment and I may immediately make another try. Two or three unsuccessful trials will usually extinguish the response temporarily, depending upon my need for the outcome and my estimate of the reliability of the response. If there is anything slightly different about the stimulus situation I am very likely to attribute the failure to that difference, sharpening the distinction between positive and negative stimuli. Attention then becomes focused on those features of the stimulus situation that are most reliably correlated with the outcome, by a mechanism that will be described later.

When no difference in sensory input can be found to correlate with success or failure we fall back on the time of occurrence as the only remaining difference. Physically identical stimuli can still differ from each other in their recency, and in general we appear to operate on the principle that the more recent outcome is the better predictor of the immediate future. As time passes, however, the influence of a few recent negative experiences are outweighed by a larger number of previous positive experiences, and the response may be repeated once more. Each unsuccessful repetition strengthens the new association, and the repetitions become more and more infrequent and finally stop.

Although little is known about the synaptic processes that result in associations, it does seem quite reasonable that they would involve a large rapid change, most of which fades away in a few minutes leaving only a slight residual effect. The residual effect accumulates, however, with repeated use and fades very slowly if at all. Alternatively some synapses may change easily and fade equally easily, and others may be changed only by many trials, but retain their modifications almost permanently.

We must assume that when two incompatible outcomes are predicted by the same response there is mutual inhibition and the stronger association suppresses the weaker. For a short time after it is formed, a recent association will dominate a more permanent one because of the strong transient synaptic change, but when this fades the more frequently confirmed association will prevail.

The inhibitory connections between reward outcome concepts and the motivational system may also be subject to extinction. It is a rather wild speculation on my part but if I may introduce a bit more physiology, I have the idea that the inhibitory terminals from the concept to the inhibitory system may require 'recharging' after use by being exposed to a substance released by the terminals of the reward path. If they are fired a number of times with no consequent firing of the reward system, they become depleted and lose their disinhibiting effect. Extinction at the concept level helps to protect the motivational associations for a time, by reducing the number of times the concept fires unsuccessfully, but eventually the connections will weaken and this will contribute to the final extinction of the response.

One of the strongest arguments in favor of *S–S* theory, as opposed to *S–R* theory, is the phenomenon of latent learning. This is learning that takes place during the exploration of a maze, or other situation, in which there is no reward. When a reward is later introduced the animal shows that it already knows its way around in the environment.

The model explains latent learning easily enough; associations are established between responses and their outcomes whether a reward is presented or not, then if an outcome is independently associated with a reward, any response predicting that outcome will disinhibit itself via the motivational connections of the reward. The fact that all contiguously active conceptual activities become associated with each other means that simple arousal of a response concept cannot be the basis for the production of an overt response. Association between *s* or *o* concepts and *r* concepts, for example, cannot be a sufficient cause for the overt expression of the response. Otherwise all responses incorrect as well as correct, that have ever been made in a particular situation may be repeated whenever that situation arises. Response concepts that are evoked by association with other concepts must still be confirmed as leading to a valuable outcome before they are allowed to reach the motor system.

I have to make this point very clear because I am now going to describe an elaboration of the basic model that will usually result in preferential excitation of the concept for the correct response in any familiar situation, and it might occur to you that the mechanism I have described earlier would thereby be made redundant.

All learned behavior can not, as most learning theorists seem to think, be fitted into the same mould. In some cases an environmental cue may provide the initiating association for a response, as in the examples considered so far, but more often behavior is triggered off by the onset of a drive, or its emergence above threshold.

So far I have mentioned drive only as a type of stimulus that allows potentially rewarding stimuli to pass through a gate to the motivation system. Hunger is necessary to make food attractive, the modulating effect of androgen is necessary to make a male interested in stimuli from a female of the species, and so on. Drive may not be the best name for this mechanism because it sounds as if it should mean the same as motivation, but I use the word to mean some state (or the neural influence of that state) that makes an animal ready to perform some consummatory sequence. System for detecting drive states like hunger, thirst, sexual arousal, sleep, and so on, have been studied in great detail by physiological psychologists in recent years and they are now fairly well understood. My interest is in following these input signals further into the nervous system to find out what part they play in the elicitation of purposive behavior.

If the only function of drive stimuli was to open gates to reward stimuli we would be unaware of our needs until some stimulus capable of satisfying one or other drive appeared, and this is manifestly not the case. We must assume, therefore, that drive stimuli, in common with others, reach the concept level and establish representative activities there. In other words we have concepts of hunger and other internal conditions.

Drive-related concepts establish associations with other concepts that are active at the same time, especially those concerned with the rewards and consummatory activity they lead to. These activities are usually going on at the same time and the associations are probably enhanced by the general arousal surrounding consummatory activity which may accelerate the synaptic changes involved.

Once associations have been formed, a developing drive state will eventually fire concepts of appropriate rewards, and possibly concepts of the responses needed to reach them. Suppose, for example, we have well-established patterns of response for dealing with excessive heat and cold; when we become overheated the concepts of a variety of outcomes, such as fewer clothes, immersion in a pool or a cool shower, standing in a breeze, and so on are fired by association from the heat regulatory drive signal. The fired outcome concepts will in turn facilitate any stimulus features that have been reliably associated with the outcomes and the responses that lead to the outcomes, and this facilitation constitutes the source of selective attention. If any feature present in the environmental input matches a concept that is being facilitated in this way from drive input, that concept will be fired more vigorously than others because of its double input. The stronger output will be reflected in stronger firing of the response concept linking that stimulus with one of the outcomes satisfying the drive (a cool drink, perhaps) and the response (reaching for the drink) will be admitted to the motor system.

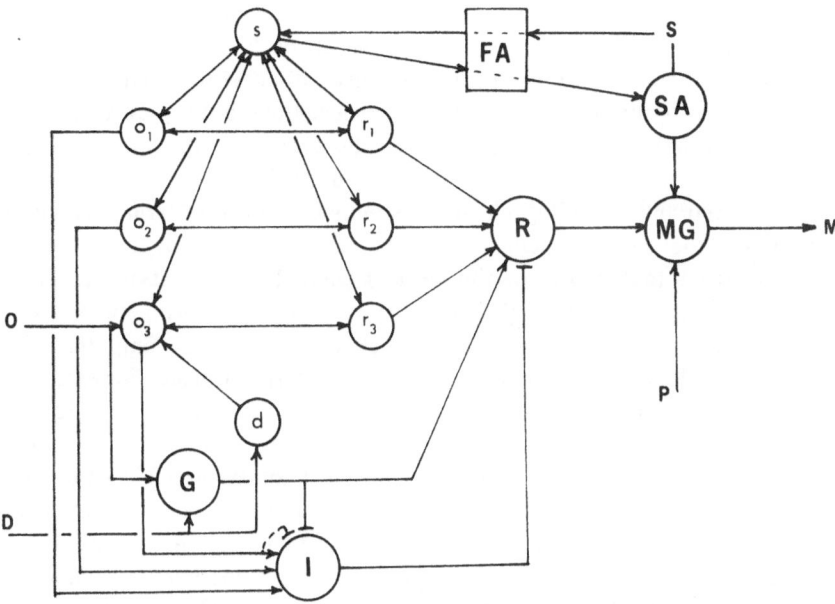

Fig. 2. A more elaborate learning model in which the conceptual influence of the drive input (d) is taken into account. FA, feature analyser; SA, selective attention gate determining which features of the stimulus should be used to guide the movements of the response; MG, movement generator in which the response instructions are converted into practical movements, taking into account the data from the stimulus and from proprioceptive in out, P. Other abbreviations as in Fig. 1.

The influence of attention does not end there, however. In many cases concepts are generalizations that do not include any of the features of stimuli that are required for the accurate direction of the movements required for responses. If I am to pick up a glass I need to know what sort of a glass it is, (especially if there are other glasses nearby) and where it is in relation to the present position of my hand. This sort of information is not likely to be given by the concept activity. The concept of a glass does not have any particular size, distance, or location in the visual field. In order to make sure that the movements of the response are directed to the right object some attentional facilitation must be directed back through the sensory system to a point where the signal still retains its spatial components and it is these signals that are enhanced, or gated, by attention to be fed to the motor system.

The somewhat more elaborate system incorporating these features is shown in Figure 2.

Part of the motivation for trying to work out this model is to give some direction to my research on the physiological mechanisms of motivation. Some people are willing to spend their lives discovering what chemicals are involved in initiating eating behavior, for example, but my own curiosity is more psychological. I would like to discover how they do it. I am particularly interested in discovering the normal function of pathways that are fired by rewarding electrical stimulation of the brain. My guess at present is that a crucial element is the path from G to I in the model.

Apart from that aspect of the work, I think I am still trying to build robots, at present only with pencil and paper, but eventually perhaps I will get back to Meccano, or its present day equivalent the computer. The technical problems are awesome, and not only a matter of scale. For example, how do you break up an $s-r-o$ loop once it is in full swing? How do you make sure that all the important responses are tested and compared before one of them is decided upon? (The problem of impulsiveness.) Then there is the whole question of concept formation including the representation of temporal sequences neurally. Many stimuli, and most responses, consist of a number of temporally distinct components, and sometimes several components run in parallel with precise timing requirements. However, nature has found ways around all these problems and perhaps we can too.

REFERENCES

Hebb, D. O.: 1949, *The Organization of Behavior*, Wiley, New York.
MacCorquodale, K. and Meehl, P. E.: 1954, 'Edward C. Tolman', in A. T. Poffenberger (ed.) *Modern Learning Theory*, Appleton-Century-Crofts, New York.
Tolman, E. C.: 1932, *Purposive Behavior in Animals and Man*, Appleton, New York.

CHAPTER 9

COMPLEXITY AND SYSTEM DESCRIPTIONS

ROBERT ROSEN

There is an enormous literature on the complexity of systems, and with attempts at specifying intrinsic measures of complexity. In this note, we will take an opposite view; that complexity is not an intrinsic property of a system, but rather manifests our capabilities to interact with the system. Since our capabilities to interact with systems around us are continually changing, so too do their apparent complexities. Some consequences of this viewpoint, bearing on system descriptions, and on the capacity of systems to make errors, will be dealt with below; a fuller discussion of these matters will be presented in another place.

Intuitively, we regard a system to be complex if we can interact with it significantly in many distinct ways, and if each of these different ways requires a different mode of description of the system to encompass it. Thus, organisms are complex; we can interact with them at many levels, and describe these interactions in many different ways. For ordinary purposes, a stone is a simple system; we typically interact with a stone in only a few ways, and basically a single mode of description is sufficient to describe these ways. If we adopt this viewpoint, that system complexity refers to our modes of interaction with the system, and to the corresponding number of different descriptions to which these modes of interaction give rise, then a number of questions immediately come to mind: (1) Is there one class of interactions, and consequently one mode of system description, from which all the others can be derived (the problem of reductionism), and (2) Is there any kind of measure of interactive capacity which will allow us to define complexity in other than subjective terms? In the subsequent discussion, we shall be concerned with some of the aspects of these questions.

As we have repeatedly emphasized elsewhere (Rosen, 1969, 1972), all system descriptions, however much they may vary in detail, ultimately consist of two basic kinds of notions; those dealing with the specification of an *instantaneous state* of a system, and those dealing with the manner in which this instantaneous state changes in time as a result of *forces* imposed on the

W. E. Hartnett (ed.), Systems: Approaches, Theories, Applications, 169–175.

system. As a typical example, we may consider the dynamical theory of particle mechanics, from which all other modes of system description have been derived. If we are given a system of N gravitating particles in ordinary 3-dimensional space, then the Newtonian formalism tells us that we can describe this system, at any instant of time, by specifying the displacements of the constituent particles from some convenient origin of co-ordinates, and by specifying their corresponding velocity or momenta. Thus, a set of $6N$ numbers suffices for a specification of the instantaneous state of the system. By this is meant that *any other* quantity pertaining to the system can be expressed as a function of the ones used to describe the instantaneous state, and hence is in a sense redundant. The displacements and momenta satisfy all the properties of a set of *state variables*; they represent a minimal set of observable quantities pertaining to the system, from which all other information pertaining to measurable quantities defined on the instantaneous states can be derived.

Newton's Laws also provide us with a dynamic description of the way in which the instantaneous states change with time, by identifying the rates of change of the momenta of the particles in our system with the forces imposed on the system. Knowing the forces, and knowing an initial state of the system, is thus sufficient to allow us to predict the state of the system at any time (at least in principle), and hence, to predict the value of any observable quantity associated with the system.

In this kind of formalism, we thus identify the abstract states of our system with a given mode of state description; i.e., with a subspace of Euclidean N-dimensional space. However, even in this formalism, there is a great deal of room for alternate descriptions. For instance, it is one of the fundamental assertions of physics that every real-valued function defined on the state space represents a potential dynamical variable of the system, and one which could be directly measured by some suitable measuring instrument. Conversely, of course, every measurable quantity pertaining to the system can be regarded as giving rise to such a function. Each such function inherits a dynamical equation, from the dynamical equations which govern the state variables. As we have shown elsewhere, however (Rosen, 1968), it can easily happen that a set of such observables will form a dynamical system in its own right; if we were to interact with the system in such a way as to observe those observables rather than the state variables, we would in effect see a completely different system. There even exist universal dynamical systems, which allow us to find observables inheriting any prescribed set of equations of motion posited in advance. Thus, even in this relatively straightforward situation, there are profound epistemological problems arising when we attempt to give an 'intrinsic' character to the results of measurements or other kinds of interactions with the system.

Another aspect of the crucial concept of interaction between systems is discernible from the fact that the state description, which presumably contains within itself the answer to every question which can be asked about that system, pertains only to a perfectly isolated system. Given two systems perfectly specified in isolation, we cannot, *from those specifications alone*, decide how those systems are going to interact. To deal with the problem of interactions, we must invoke further information, usually in the form of some universal principle (e.g., mass action) which never forms part of the state description of individual systems. This is the primary reason why physics is successful in dealing with isolated systems, but cannot cope with inter-actions, except in the case in which they are infinitely weak (perturbations).

We have thus seen (a) that even a 'simple' physical system can be interacted with in many ways (i.e., through many different sets of observables of the system); each of these modes of interaction conveys some aspect of reality pertaining to the system, but the system will 'look different' from mode to mode; and (b) the capacity of a system to manifest interaction is never considered a part of physical system descriptions, although it is crucial for such areas as biology. There is no reason why we should expect that two interacting systems will 'see' each other through the same observables through which we find it convenient to interact with them separately, and consequently, there is no reason to expect that our conventional modes of state description will allow us to understand (let alone predict) the results of the interaction.

The above remarks bear heavily on the problem of reductionism, and on the more general problem of approaching complex systems through analytic means. By *analysis* we mean here the resolution of a system into a family of subsystems somehow 'simpler' than the original system from which they were extracted, and attempting to infer the properties of the original system from the properties of the subsystems. The extraction of a subsystem corresponds formally to a process of *abstraction*, in which a number of degrees of freedom of the original system (i.e., potential interactive capabilities) are excluded, and only a limited number are retained. This process of abstraction can be physically implemented (as when a molecular biologist extracts a fraction of molecular species from a cell, thereby creating an abstract cell) or they can be purely formal (as when an ecologist represents a population of real organisms in terms of predation relations). The basic requirements of such abstractions are the following:

(1) The subsystems so obtained must be 'simpler' than the original system from which they were abstracted;

(2) The subsystems must be obtained by 'natural' means (i.e., utilizing familiar and justifiable procedures); and

(3) The properties of the subsystems so obtained must permit the determination of the properties of the original system.

The property (1) is obviously crucial; nothing is gained if we extract systems as intractable as the original system. This has long been recognized implicitly in scientific modes of analysis. Of equal importance is the property (3); any property of isolated subsystems not bearing on the properties of the original system is an *artifact*. The property (2), however, is a purely subjective matter, and refers only to the manner in which we find it convenient to interact with the original system. It thus stands on a different footing from (1) and (3).

Nevertheless, in many empirical modes of system analysis, the greatest weight in placed upon condition (2). It seems to be intuitively hoped that, by relying on procedures which satisfy (2), the conditions (1) and (3) will automatically be satisfied. At the very least, it is hoped that (1) + (2) will imply (3). However, from what we have already said, this is plainly absurd, in general. Indeed, what we learn from the above is that the crucial properties (1) and (3) which must be satisfied by any useful means of analysis of systems, must be allowed to determine what we are to regard as 'natural'. Indeed, 'naturality' must not be allowed to be posited in advance, but only in terms of its bearing on the problems under discussion in a particular context.

A simple example may make this clear. In physics, the three-body problem is complex in a well-defined sense; the dynamical equations governing a system of three gravitating masses in an arbitrary configuration cannot be integrated directly. We could hope to approach this kind of problem by analysis into a family of 'simpler' subsystems, which will allow us to solve the problem. Intuitively, the subsystems available to us are two-body systems and one-body systems. These are indeed 'simpler' than the original system, and are abstracted from that system in 'natural' ways. However, it is clear that we cannot solve a three-body problem in this fasion, for the act of decomposing the original system into isolated simpler subsystems destroys irreversibly the dynamics in which we were originally interested (here again we see the inability of physics to deal with arbitrary interactions). Thus, from the standpoint of solving the three-body problem, our apparently 'natural' decompositions are useless; if analysis is to be successful in this kind of problem at all, the appropriate subsystems (i.e., those which satisfy (3)) must necessarily be of a kind which would appear most 'unnatural' in terms of what we find it convenient to do physically to a system of particles.

An abstraction, in the sense in which we are using the term, owes its efficacy to the removal of degrees of freedom (i.e., interactive capabilities) present in the original system. It represents a simplification in that it faithfully represents what the original system would be like if it could only interact with the world through those degrees of freedom which are retained in the abstraction. Thus, if the abstraction exhibits similar interactive capabilities to those possessed by the original system, we are justified in assigning them to the properties retained in the abstraction. (However, we must note that an abstracted subsystem may exhibit new interactive capacities of its own, which do not pertain to the fact that it is in fact a subsystem of some larger system; always carefully distinguish between those properties which inhere in the original system, and those which arise in the model without reference to the fact that it is a model.)

Let us suppose that we have two different abstractions of the same complex system. How may these be compared or combined to give a fuller picture of the original system than either of them can given alone? We shall touch briefly on this question, and show how it relates to the notion of *structural stability*, which is presently attracting great interest in theoretical biology and elsewhere, largely through the influence of Thom (1972). Indeed, structural stability is a general framework for comparing one description of a class of objects against another description, usually in terms of measures of 'closeness'. Typically, if two objects 'close' in one description are also 'close' in the other, then the objects are called *generic*; generic objects are thus precisely those for which the two descriptions agree. Those objects which are not generic are called *bifurcation* objects; objects 'close' to a bifurcation object in one description will not be 'close' to that object in the other description. Generic objects thus do not require both descriptions; the descriptions are redundant on these objects. On the bifurcation objects, however, the two descriptions are conveying different information about the objects.

It is usual, in considerations of structural stability, to regard one of the two descriptions being compared as *intrinsic*; the second description is then compared with the intrinsic one. However, as we repeatedly emphasized, (cf. Rosen, 1973) the topologies (i.e., the measures of 'closeness') which we typically assign to systems and their states are not intrinsic, but are all contingent on how we choose to interact with a system. Thus, any description is at best conveying partial information about a limited fraction of the interactive capabilities of the system, and all are equally extrinsic.

Nevertheless, if we have an idea that a particular description is intrinsic, and that all other descriptions should be referred to it, we will typically find some bifurcation set where the two descriptions do not agree. Since we tend to prefer a description we regard as intrinsic, it would be reasonable for us to interpret the disagreement between the descriptions as a *mistake* or *error* on the part of the second description. Thus it is at this point in the analysis of complex systems that the concept of error arises in a natural way.

Error has always been troublesome for system theorists. Simple systems do not make errors; it is meaningless to regard a system of mechanical particles as behaving erroneously. Therefore there has always been a relation between complexity and the capability or error; just as with complexity, there has been an enormous literature generated in an attempt to obtain an intrinsic definition of error. We will argue, analogous to our argument regarding complexity itself, that error is not definable intrinsically, but only in terms of the spectrum of interactions available to a complex system. We shall also find that the conventional view, that error depends on complexity, is well-founded.

To put the case simply, we shall assert that error is measured by the deviation observed between the actual behavior of a complex system (with its many distinct interactive capabilities) and the behavior of a simple system or model, which exhibits only a restricted subfraction of the interactive capacities of the original system. The deviation between these behaviors arises precisely because the complex system can be, and in general is, doing many different things at once, and these different interactions interfere with each other in a way unpredictable in principle from the properties of the corresponding simplified system (since, of course, the vehicles for these interactions have been abstracted away, and hence permanently lost, in the very process of abstraction which led to the simplified system). Thus, a bridge can collapse because the particles of which it is composed may interact with each other in ways not consistent with the maintenance of cohesive properties; the genes of an organism can mutate because they can interact with elements of their environment in ways not compatible with their coding function, and so on. I would argue that, in fact, all forms of what we interpret as error arise through simultaneous manifestation of modes of interaction in complex systems, which happen to interfere with each other.

These observations have many implications for important areas of control theory and error correction at the technological and social levels, as well as for an understanding of biological and physical systems. For example, we may compare a model of our own regarding a particular biological and social process (i.e., a simple representation of a complex system) with that manifested by natural selection acting on that complex system. A comparison of the two descriptions will reveal bifurcation points, at which the selection behavior manifestly disagrees with our projections for system behavior. These are the points at which great care must be taken in exercising control. For instance, if we decide we want a system to behave optimally, or without error, in a single simplified mode, we can only achieve this at the cost of interfering in unpredictable ways with other modes of system interaction. We have discussed these problems, in a variety of contexts, in other work, to which we refer the reader for fuller discussions (Rosen, 1974). See also the papers of Pattee (1971). Further, it is important to note that it is in general not sufficient to attempt to replace the missing interactive capabilities which were abstracted away in the initial simplification of the system, by some generalized probability distribution imposed *ad hoc* on the degrees of freedom remaining in the simplified system. This is, of course, the usual procedure adopted in this regard, but from a more general perspective it is fraught with pitfalls.

We hope to have indicated, in the above brief survey, the epistemic character of the notions of system complexity, and the related problems of alternate description and of error in complex system behavior. These notions, we feel, are of great importance in attempting to deal with the spectrum of technological, social and biological problems which presently confront us.

ACKNOWLEDGEMENT

This paper was prepared with the support of NIH grants #2RO1 HD05136-04 and #1PO1 HD07328-01, and NASA Grant #NGR33015002.

REFERENCES

Pattee, H.: 1971, in R. Buvet and C. Ponnamperuma (eds.), *Chemical Evolution and the Origin of Life*, North Holland: Amsterdam.
Pattee, H.: 1976, *Biogenesis: Evolution and Homeostasis*, Springer-Verlag: Heidelberg and New York.
Rosen, R.: 1968, *Bull. Math. Biophys.* 30, 481–492.
Rosen, R.: 1969, in Saaty and Wegl (eds.), *Mathematics in the Sciences*, McGraw-Hill, New York.
Rosen, R.: 1970, *Dynamical System Theory in Biology*, Wiley-Interscience, New York.
Rosen, R.: 1973, *International Journal Systems Science*, Vol. 4, No. 1, 65–75.
Rosen, R.: 1974, *International Journal General Systems*, Vol. 1, 245–252.
Rosen, R.: 1974, *International Journal General Systems*, Vol. 1, No. 2, 93–103.
Thom, R.: 1972, *Stabilité Structurelle et Morphogénèse*, W. A. Benjamin, Inc. Reading, Mass.

CHAPTER 10

CONCERNS, COMMENTS, AND SUGGESTIONS

WILLIAM E. HARTNETT

1. EDUCATIONAL CONCERNS

In this paper, I want to talk about some aspects of Education which are frequently ignored. For the most part, I shall limit myself to discussions involving Science and Mathematics, although I feel that the ideas have wider applicability.

Progress in Science and Mathematics depends on discovery and/or creativity. But it also depends very heavily upon education and one can make a strong case that the weight of traditional views frequently impedes progress when transmitted to the next generation. One might even argue that we presently impede progress because we do not take education seriously. One of my friends remarked: "Doing Physics is acceptable but examining the foundations of Physics is thought to be a second-rate activity". With such a criterion, what I am concerned with is probably no better than sixth-rate in the view of some. Because I am convinced that there are always a certain number of misguided persons who will remain in that state, I shall not attempt to mount any justifying argument for my interests or concerns. Instead, I shall turn to them immediately.

The most striking problem in Education is a need — the need for new (and better) ways to learn. Few of us are satisfied with our own education — I think that mine, in many ways, was dreadful — but the education of present-day students does not seem to be all that much better. At a time when there is so much to learn and when so much needs to be done, we seem to be burdened with unsatisfactory ways to learn. And it is obvious here that if we do not gain ground, then we certainly lose it. I feel that we are not now gaining ground.

Granted that there is a need, what can we do? Anything? Yes, of course, there is. There are various answers which range all the way from changing from a red (but unread) textbook to a blue textbook to doing 'real problems' to using a self-paced instruction. These answers, as far as I can see, fail

W. E. Hartnett (ed.), Systems: Approaches, Theories, Applications, 177–197.
Copyright © 1977 *by D. Reidel Publishing Company, Dordrecht-Holland.*
All Rights Reserved.

to provide what we need. There are other answers, however, which can be given which have the possibility of meeting the need for fundamentally new ways to learn. In the case of Mathematics, some of these have been written down in some detail (Hartnett, 1963, 1971, 1973). In this paper, I will deal with notions which should help with some of the problems of science education.

Earlier on I talked about the need to fashion new ways to learn. Those not already convinced of this have only to look at the usual undergraduate or graduate student, specializing in a particular field. What one sees can best be described as a 'possessor of bits and pieces', a person armed with scraps and fragments of knowledge, desperately trying to remember them in case some professor demands instant retrieval. I would argue that learning would be improved if we could do something about the 'bits and pieces'.

The first thing we might look for are ways to unify, even at the expense of oversimplification. What the particular ways to unify are will, in part, depend upon the field, but they will also depend on those involved with the education of students. These latter include the usual classroom teachers and the writers of texts and treatises. In my experience, few of them make any attempt to unify and so we continue to produce 'possessors of bits and pieces'. Much could be done on the question of unification.

Akin to unification, but somewhat different from it, is the question of a conceptual framework. Again at the risk of oversimplification, a conceptual framework provides a student with a place to hang his 'bits and pieces' so that he can have some understanding of the relationships between them. The only framework the student presently has is the course reference − "that was covered in QED 102 that I took last year − I think". If we persist in chopping learning (or is it knowledge that is being fragmented?) into courses, is it any wonder that courses provide the only framework they have? While there will invariably be arguments about which conceptual frameworks are worth-while, *some* conceptual framework can be invaluable for a student. Any such framework can be modified as needed and still remain useful. Of course, the adoption of a particular framework may force substantial changes in the way one carries out certain aspects of his educational efforts. We look at an example. Suppose we set a conceptual framework for Physics by asserting that Physics is the study of physical change. Then as one deals with various traditional concerns of Physics one always formulates matters in terms of physical change. This leads to standard questions: What changes? What stays the same? How does change occur? Why does change occur? And on and on. Another example: a conceptual framework for Mathematics is expressed in the view that Mathematics is the study of sets with structure (the 'sets and functions' approach) and those (standard and otherwise) questions which can be asked when dealing with sets having structure.

If either of these conceptual frameworks were actually given widespread currency, then we would certainly have to change the way we are teaching today. (Parenthetically, the suggestion of a conceptual framework for a student is only providing what the teacher presumably already has and uses. The claim here is the student could also use it profitably even from the beginning of his work.) We could no longer endure – nor would the students tolerate – the spectacle of bits and pieces strewn with wild abandon around the intellectual landscape. They would ask the now obvious questions: do these bits and pieces fit into the framework? If so, then how? If not, then why not?

Another possibility as a new way of learning is the use of organizing principles. As in the case of a conceptual framework, the experienced learner (= teacher) has come to certain organizing principles, perhaps painfully so, by a series of disjointed experiences. However, with rare exceptions, he makes little attempt to organize his teaching so that his students may be lead to these principles quickly and efficiently. I would be the first to assert that organizing principles cannot be taught directly. However, it is possible to select what one does and how one does it with an eye to having the student acquire the principles. (A more extensive discussion of a 'principles' approach coupled with a 'conceptual framework' approach, in the case of Mathematics, can be found in Hartnett (1973).) Once again, a 'principles' approach, if implemented, would have a profound effect on what and how we teach. Think what would happen to our omnipresent (and largely useless) calculus courses! Not to mention the profusion of calculus books.

A fourth possibility for a new way of learning is generalization and the general systems approach is one example of this way. Unfortunately, this way seems to require enough concrete instances before one can successfully hope to achieve the appropriate level of sophistication, especially if the approach is heavily mathematical. Despite this drawback for the beginning learner, this way of learning has been important and will likely become more important in the future.

Having now suggested four possible ways of learning (not all new but certainly all under-utilized), it would be fitting if we could spell something with U, CF, OP, and G and raise high a banner in the breeze which bears a slogan for the campaign. It might be fitting – even proper – but we will put it aside for a moment to get on with the work at hand. What I specifically want to do in this paper is to present some ideas about mathematical models, especially for Science. What I shall suggest is a conceptual framework for mathematical models, embodying some organizing principles, providing unification of concepts, and generalizing a variety of concrete situations in modeling. I shall place this in a general context, but shall give specific illustrations only in the case of Physics. Throughout the paper, I shall be more suggestive than specific. Much of what I contend here needs amplification and justification. Both will appear in a future book on these matters.

2. USEFUL MATHEMATICAL MODELS

The use of Mathematics in some fields has a long history — Physics is surely the prototype utilization of Mathematics. But the explicit notion of a mathematical model in the sense in which I want to use it seems to be a product of this century. It has only been in very recent times that much effort has been expended on a serious use of mathematical models. There are few texts on general mathematical modeling and those which do exist have serious shortcomings. It would be pleasant if one could formulate notions about mathematical models which would: (1) be useful for beginning students in terms of UCFOPG, (2) be available for use in the 'hard' sciences, the 'soft' sciences, and system sciences, and (3) be of such a nature that it could be of use in the creation of an exact (= mathematical) philosophy. In essence, that is what I would like to accomplish. Whether I do is another matter. The best I can really hope for here is a sketch of the ideas involved. But if the ideas are used, then we will have to make substantial changes in what and how we teach. In particular, we will have to achieve a much greater clarity and precision than we have in the past. Such clarity and precision would, of themselves, be of considerable aid as a basis for UCFOPG.

3. PROBLEMS OF APPLIED MATHEMATICS

Mathematical models have already been used with great success in fields like Physics, Chemistry, and Engineering. It is likely that their use will become more widespread in other fields. Even so, I do not feel that the models or their uses are well understood. If we are to make progress by using mathematical models, then we must know what we are doing. It is not enough to just do. Again, a problem of Education.

Quite apart from a conceptual framework for mathematical modeling, there are at least two other matters to attend to. Clearly, mathematical modeling involves applications of Mathematics. Hence one may ask: what Mathematics? The question is not quite trivial. As mathematicians know, there has been an absolutely spectacular development of Mathematics in this century, especially in the last twenty-five years. We now have a 'warehouse', filled to overflowing with mathematical results and techniques and much of the stock of the warehouse is new and exciting. All of it is available, free of charge for the taking (and using). But our possessors of 'bits and pieces' are still learning the same old ways and the same old results. In particular, they are, for the most part, learning only those uses of Mathematics which their teachers know. But the warehouse is building new annexes every day to accommodate the new material that is arriving in a steady stream. The problem then becomes a question of access for new learners. Their teachers know only very little of what is in the warehouse. What they use, they learned from their teachers. In the past, the warehouse was underutilized. But now the problem is even more serious. Much of the new material in the warehouse is written in a different language and from a new viewpoint. There appears to be no way to use what is there unless we find new ways to learn it — and unless we learn the language needed to gain effective entrance to the warehouse. (The language, by the way, is sets, functions, and relations. Our earlier suggested conceptual framework is that Mathematics can be 'said' in sets and functions.) Because the question of the warehouse is of critical importance in applications of Mathematics we list it as an explicit problem.

PROBLEM 1 (of Applied Mathematics): Find ways to utilize the warehouse!

The second item to deal with is, in a sense, the reverse of the first. It has to do with the mindless festooning of everything in sight with Mathematics or at least with the trappings of Mathematics. This is painfully apparent in the literature. Papers are written and advertised as having to do with mathematical modeling. Quite frequently, it is not clear what the problem is (or indeed that there is a problem), it is not clear what is being modeled (and how), and it is not even clear what the author hopes to accomplish. I am vitally interested in the applications of Mathematics for a variety of reasons but I am only interested in reasonable applications. Much of what I see is not reasonable. So we identify another problem.

PROBLEM 2 (of Applied Mathematics): Find ways to apply Mathematics reasonably!

Apropos the second problem, there seems to be the feeling abroad that "if you're doing something, then you must be doing something worthwhile". It is reminiscent of the early days of computers. Then the status symbol was the computer printout. Now the claim is that "we have quantified the whole approach". Perhaps it is time for a new slogan to replace GIGO (= garbage in, garbage out).

Applied Mathematics is productive of many fascinating questions, some of which are discussed in Nalimov (1974), but none of which can be treated here.

4. MODELING

Man has always tried to understand what goes on around him. His attempts have ranged from avoidance of danger – "big bears bite so don't fool around with bears" – to placation of the gods (known and unknown) – "they may not matter but, then again, they might" – to the complex, complicated, and sophisticated activity we call Science.

In earlier times, it was relatively easy to say what Science was. It may never have been easy, but, in any event, it is harder to say now. However, almost all would agree that Science is an activity which involves the study of something. Much argument can and does ensue about exactly what the activity is, how it is carried out, and what the something studied is. This paper tries to provide a discussable framework within which one can try to deal with some of these questions and which may permit the formulation of questions not yet considered.

The specific focus of the paper is on the use of mathematical models in this general framework and, by specialization, in Science. Ideally the paper should be burdened with examples – reams of examples! By then, however, we would have a book not a paper and, possibly, a much smaller audience. So, to popularize the framework and approach, we forgo the grand numerosity of examples and settle for only a few.

Motivation for the paper comes from watching scientists try to use mathematical models. Science is hard work and should not be made unduly difficult by the use of mathematical models. If the mathematician interested in Science can do anything "to ease the pain", then, of course, he should. And so I shall try. Recall, however, that much of my interest is concerned with the beginning learner, not with the active, mature scientist.

I hold that there are things without committing myself to a detailed explanation of what things are. Some things are more thing than other things. This paper is a thing, this pen is a thing, and this desk is a thing. Other things I know are also things. However, I am less confident about reported things in intergalactic space, for example, black holes. Generally speaking, one is somewhat limited to 'nearby' things although extensions are possible.

Molecules are things, as are cells. When organized into organisms we have things made up, in some sense, of other things, albeit in a mystifying and not yet understandable way. The point to be made is that there are different kinds of things ranging from blocks of granite – sturdy but not fascinating, except perhaps to the geologist or the sculptor – to persons – often not sturdy but nonetheless always fascinating. Suffice to say for now we want to deal with interesting things. Actually, we want to talk about special interesting things, not all interesting things.

Things are not interesting unless they are doing something we can observe and even then we tend not to pay too much attention to them unless we are puzzled or intrigued by what they are doing or how they interact with us. Science is the thinking man's answer to puzzle or intrigue. It involves a serious study of the puzzling or intriguing things with a view to explaining what is happening, what has happened, and what may happen. Some things do not seem to be studiable – angels are not studiable, at least not in Science – and perhaps some things are not worth studying. I would, however, like to argue that when we study studiable things we do so in certain ways which involve what (for want of a better word) I call *schemes of things*. In this and subsequent sections, I will amplify this notion and use it to discuss mathematical models. As I indicated above, I do not want to make the notion of thing precise and the same could be said for the notion of a scheme. The reason for this should be mentioned – I do not want to unduly limit this conceptual framework involving mathematical models. We will discuss schemes, however, and look at some examples.

It is convenient to coin the term *object* to denote a studiable thing, the implication being that objects are interesting because they are doing something puzzling and/or intriguing. There is a classical situation which we can use to discuss the notion of a scheme – the situation of the police inspector (name your favorite sleuth) confronted with the crime or crimes. He has certain facts, certain clues, certain phenomena – body found in a locked room, etc. Question: Who did it? Answer: Think up a plausible scheme which explains how it *could* have happened, make deductions from the scheme, make predictions about the possible response to actions, and finally "try it on for size". Next step, new questions: What do I know now that I didn't know before; has anyone become a more likely suspect? And so it goes: first, the phenomena to be explained, second, the plausible scheme, third, the trial by fire for the scheme. Next step, next questions: How did the scheme fare; did it fall apart; did it have to be refined; was it worth anything? Depending on the answers, one cycles back to the scheme for a refurbishing, arrests the now obvious murderer, or pours a drink and starts all over again.

In this case, the phenomena leads to an explanatory scheme. Pictorially we have:

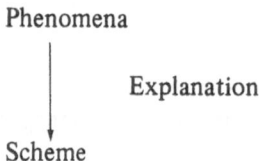

As indicated before, a first scheme might lead back to new phenomena which then leads to a new scheme, etc. Schematically, one would have:

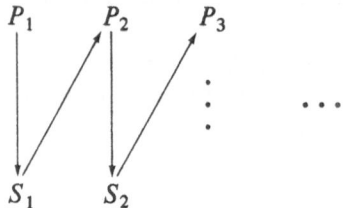

For simplicity, we use only the first diagram above with the understanding that it may involve something like the second diagram.

It should be noted that our detective inspector seeks an EXPLANATION but it must be consistent with what facts he has and with what he knows about a body, locked rooms, and murder weapons. In short, he seeks an explanatory scheme within a CONTEXT which determines the consistency of his scheme.

Shift now to a different example — one from radio astronomy. One dark night, Jodrell Bank picks up a repeating, pulsing signal. Standard response: some sort of noise from within the observatory or from nearby installations — find it. After a long search for the noise source, the conclusion is reached that the signal is coming from 'out there' and the obvious questions are now asked: who or what could be sending out such an unusual signal; what is going on 'out there'? The current answer is that we are listening to pulsars and the pulsar scheme includes the result of observations, the invention of a celestial system mechanism to explain how the system makes that peculiar 'noise' and finally an explanation of how such a system could come into being. The first, generally speaking, deals with facts, although questions of our observational influence could be raised. The second must not violate what we know (or at least believe) about celestial systems and should be testable by observations and/or experiments. Finally, the third must be compatible with our view of the origin of celestial systems. Again we have puzzling phenomena and an explanatory scheme within a context.

In the particular case of pulsars, one can trace the formulation of (and rejection of) various explanatory schemes and one presumes that more schemes will arise in the future. The cycle is: phenomena, scheme formulation, scheme manipulation, scheme testing, scheme evaluation, scheme revision – all this interspersed with observations as appropriate. Naturally, the cycle is not carried out entirely by one person; the task (if serious) is too great. In practice, many persons join in the task which may stretch over a number of years and which rarely terminates with any finality.

If we have created a scheme for an object (= a studiable thing), we generally want to manipulate the scheme either to find out more about the object or about the worth of the scheme itself. Because a scheme is an explanation of what we have observed, we might first see what this explanation entails; all too often it requires the impossible and hence we must modify it or discard it. (Think about our master detective!) If we finally fashion a decent scheme with no obviously disastrous logical consequences, we can then begin to explore what possible predictions could be made from the scheme. Here again we may come up with predictive situations which we could never create or at least not create without staggering effort and hence we would not have any testable predictions. This would mean that we have to work harder at producing manageable predictions or go back and start over with a new scheme. Hence the decent scheme should allow, permit, or even encourage

(1) derivation of consequences with some ease, and

(2) production of predictive situations.

Now for many phenomena one can fashion suitable explanatory schemes which have properties (1) and (2) and hence one can successfully deal with the phenomena via the manipulation of the suitable scheme. In a number of other situations, however, it is difficult, if not impossible, to manipulate the scheme and one then tries other avenues. Depending on what one is trying to do, that is, how and what one is studying, one chooses a representation which is amenable to manipulation. These may range from an actual scale model to be used in a wind tunnel to a computer simulation model to a mathematical model. (One should note in passing that 'model' is frequently used for what I have called 'scheme'. I prefer the present terminology because it allows us to assert that mathematical models do indeed model something – they model schemes of objects. The beginner might find such a viewpoint helpful.) Our concern here will be with mathematical models, which is not to say that the other kinds of models are not useful. They have been and they are.

To move from a scheme to a mathematical model, one must represent everything in the scheme in the world of mathematics, that is, one must interpret the elements of the scheme as mathematical entities. The history of Physical Mechanics, both Classical and Quantum, demonstrates that there are no God-given rules which tell you which things should be interpreted as which mathematical entities. After all, the warehouse is full today but will be even bigger tomorrow. No rigid and eternal rules could deal with this growth of potentiality. There is so much to choose from!

Many persons, who build and utilize mathematical models, feel that a mathematical model should 'mirror' what is 'really happening'. It is not completely obvious that such a view is appropriate for the progress of Science. Indeed, it seems unduly restrictive. After all, one's real concern in studying is to be able to learn something. And it just may be that one can learn by utilizing new and perhaps mysterious mathematical entities. (In the early days of Quantum Mechanics, it was not clear that certain Hermitean operators on Hilbert space were 'mirroring' the observables of physical systems. The situation now is even less clear.) Here again, we have the beginning learners learning only what their teachers know. For the most part, we don't suggest that they think of alternatives; this is because they don't even know what alternatives there might be and because they have no effective way to get into the warehouse. This has to be changed.

We look at an example – the classical mechanics of a physical particle, presumably a mathematical model of motion. In usual terms, motion involves change of location, location means a coordinate system, and change means differential equations. Indeed, the scheme gets little attention here, the scheme essentially becomes the model from the beginning. In the case of Mechanics this particular approach has been remarkably successful although, as Wigner (1960) notes, it has no business being so. However, the very success of this approach may have ended up giving a bad example. Too many people in other fields, smelling the sweet aroma of borrowed 'success', rushed to mathematize only to discover that they didn't really know what they were mathematizing or what they were modeling. Had they spent more time formulating and modifying a suitable explanatory scheme, they would have more to show for their efforts.

To be more explicit, consider the case of a single physical particle, moving in a straight line. In the usual treatment, one talks about the mass of the particle and, at each moment of time, the position or location, the change of position (= velocity), and the change of velocity (= acceleration) of the particle. One also talks about possible forces acting on the particle. Finally, one makes some statements about the relationships between the notions involved, for example, "force causes a change of position" and "more force causes more acceleration".

Interpreted, time is represented by $\mathbb{R}^+$, the set of non-negative real numbers, mass by a function $m: \mathbb{R}^+ \to \mathbb{R}^+$ (usually taken to be a constant function), location by a function $x: \mathbb{R}^+ \to \mathbb{R}$ (with $\mathbb{R}$ embodying the notion of an origin $0 \in \mathbb{R}$ and a direction $1 \in \mathbb{R}$, that is, a coordinate system), velocity by a function $v: \mathbb{R}^+ \to \mathbb{R}$ such that $\mathbf{D}x(t) = v(t)$ for each $t \in \mathbb{R}^+$, acceleration by a function $a: \mathbb{R}^+ \to \mathbb{R}$ such that $\mathbf{D}v(t) = a(t)$ for each $t \in \mathbb{R}^+$, and force by a function $\mathbf{F}: \mathbb{R}^+ \to \mathbb{R}$. The statement of a fundamental observation (or requirement) involving motion is that $m a(t) = \mathbf{F}(t)$ for each $t \in \mathbb{R}^+$. If we agree that the 'state' of the particle is the information, (where the particle is, what it is doing), then the two functions x and v represent the state of the particle, and we then have a triple of mathematical objects (state function(s), change condition(s), law(s)). The state functions are x and v, the change conditions are $\mathbf{D}x = v$ and $\mathbf{D}v = a$, and the law is $m a = \mathbf{F}$, Newton's second law of motion. Such a triple is an instance of what I shall call *a dynamical system*, in this case, *for the particle*. It is a mathematical model in the world of mathematics, obtained from the scheme via interpretation.

Pictorially we have

After we have our dynamical system (DS), we want to transform it in order to learn something or do something. In our case of the particle, the actual state functions x and v are not known. The aim is to obtain them from the law. The idea is that force causes motion and so if we know **F** and the law (the *how* it is caused), then we should be able to produce x and v. Technically **F** should yield v and v should yield x. Normally, one assumes that **F** is pleasant enough so that, given the law, one can obtain the state functions. In general terms, we transform the dynamical systems in an appropriate fashion.

Returning for a moment to our general notion of a scheme, we recall that a scheme is basically a contextual explanation. If we formulate a scheme, then we might want to check it against a particular phenomena of the kind that the scheme is supposed to explain. I shall call that particular phenomena *a realization*. For example, suppose that our scheme is intended to explain how automobile engines operate in particular circumstances. A realization is then a specific engine and we ask whether our scheme successfully explains the operation of this engine in the particular circumstances. Generally, our scheme predicts something about the operation of the engine and our examination of one or more realizations allows verification of the predictions.

Success at verification is encouraging but failure at verification is fatal. Failure means we have to go back and look at our scheme, perhaps modify it, look at realizations, and, in general, try again. The roads between schemes and realizations are always well-traveled.

In the special case in which we move from the scheme to DS and then to the transformed DS, we hope that the transformed DS will allow us to make predictions about realizations and hence verifications of the scheme via its DS. We show matters pictorially:

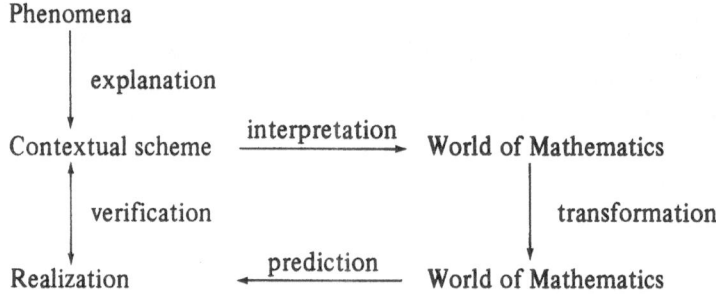

The picture is a conceptual framework for mathematical modeling and is, I claim, useful for a variety of persons. It is useful for beginners, although it will bring up some embarrassing questions, and may even be for others who have always been just a shade uneasy about the whole business of mathematical modeling.

There is one danger here that I would like to comment about because of possible misunderstanding. I intend this conceptual framework to be as general as possible — in a more extended setting, I would be able to give many examples to indicate the scope involved. Notice that I have avoided precise definitions of things and phenomena. I want to include the usual physical phenomena but also include behavioral phenomena. In particular, the phenomena might be persons thinking, the scheme explains how, etc. In other instances, the phenomena may be specific results within a discipline, the scheme then attempts to explain how these results occur. Certainly, the phenomena-to-scheme link should have no explicit restrictions on it. Earlier, I indicated some misgivings about the mindless rush to mathematize but, barring such activities, I would allow any interpretations in the World of Mathematics, our old warehouse with limited access. Of course, the point of going to a mathematical model is to gain something, perhaps precision of expression, perhaps manipulatability, perhaps predictability, perhaps all three. It is good to know what you want before you go shopping.

The reader is invited to fit his favorite mathematical models into the anti-Procrustean conceptual framework sketched above. With enough pushing and shoving, it should expand or contract enough to accommodate them.

I would now like to turn to the more general task of talking about state models of objects in a fashion which may be helpful for both scientific and philosophical considerations.

5. STATE MODELING OF OBJECTS

As far we can determine, objects 'do something', they change in discernible ways; possibly some things don't do much but objects certainly should and indeed they do. If we observe an object, we can note and record various properties that it has when we begin to observe and note and record changes in these properties as we continue to observe. If we identify the list of properties at a time with what the object is 'doing' at that time, that is, with the state of the object at that time, then we could agree that the state of the object at a particular time is the list of the properties observed at that time. Note carefully that this says nothing at all about which properties the object indeed *does* have, only about certain properties observed at that time. From this point of view, the state of an object is, in some sense, determined by the observer from a collection of possible states of the object.

It is clear that a scheme of an object, along with dealing with (1) and (2), must incorporate the notion of the state of the object at some time of observation. If we elect to gain manipulative facility by forming a mathematical model of our scheme, then our model must reflect the 'state' nature of the scheme and hence we must formulate the idea of a state model of the object.

Our problem then seems to come to this: devise a mathematical model for schemes which incorporates (via interpretation) the notion of state for objects in such a way that specialization yields the classical notion of state and which is such that the model allows the greatest possible generality.

I would like to suggest that, in a suitably general sense, all schemes of 'change' be modeled by dynamical systems. These dynamical systems involve state functions (of the objects of the scheme), change conditions, and laws. As usual, I would like to insist on as much generality as we can muster for all three parts of the dynamical systems. One might even suggest that all schemes are schemes of 'change' because the only interesting activity we observe is change. This assertion could be argued in detail but will not be so argued here.

I now set down the formal development of state functions. As we go along, I will interpret or comment as appropriate.

For the moment I proceed as follows. Assume that certain objects may be associated with certain undefined and, for now, unspecified functions called *state functions*. If θ is an object, then θ is *stateable* iff θ has a state function associated with it. If f is a state function for stateable θ, then the codomain of f is *the state space of θ for f* and for each $x \in \mathscr{D}_f$, $f(x)$ is *the state of θ at x (for f)*. The range $\mathscr{R}_f$ is *the lawful state space of θ for f*. Clearly, all of these notions depend solely on the state function of θ that we have at hand. (It would be possible to deal with mathematical objects more general than functions, for example, the morphisms of Category Theory. It is not clear, at the moment, what the gain would be.)

Interpretation. Objects are studiable things. Such things have properties. Each element of the domain of a state function of the object (= a 'time' set or a 'time' set together with 'input' which might be past 'history') is regarded as the 'situation' of the object at some instant or time and its image under the state functions 'measures' the properties of the object in that situation.

It should be explicitly pointed out that I place no restrictions on the domain and codomains of state functions of stateable objects. In particular, I assume no algebraic, topological, or analytic structure. Nevertheless it is possible to suggest reasonable tentative axioms for state functions. Relatively modest, the axioms generally tell how to get new state functions from old ones. I propose the following for a fixed stateable object, θ.

Axiom A. If $\mathbf{f}$ is a state function for θ, then card $(\mathscr{D}_f) \geqslant \aleph_0$.

Comment. If we are going to model change in objects, then we should allow ourselves 'enough' time for the changes to occur. See Windeknecht (1971).

Axiom B. If $\mathbf{f}\colon T \to X$ is a state function for θ and $\mathbf{h}\colon X \to Y$ is a state function for θ, then $\mathbf{h} \cdot \mathbf{f}$ is a state function for θ. (In particular, if $T = X = Y$, then all state functions for θ in T^T form a monoid under composition.)

Comment. A 'measure' of a 'measure' should again be a 'measure', although it may not be a satisfactory 'measure', in any sense at all.

Axiom C. If $\mathbf{f}_i\colon T \to X_i$ is a state function for θ for $i = 1, 2, ..., n$, then the function $(\mathbf{f}_1, \mathbf{f}_2, ..., \mathbf{f}_n)$ defined by

$$(\mathbf{f}_1, \mathbf{f}_2, ..., \mathbf{f}_n)\colon T \to X_1 \times X_2 \times ... \times X_n$$

$$t \in T \xrightarrow{\;(\mathbf{f}_1, \mathbf{f}_2, ..., \mathbf{f}_n)\;} (\mathbf{f}_1(t), \mathbf{f}_2(t), ..., \mathbf{f}_n(t))$$

is a state function for θ.

Comment. If each $\mathbf{f}_i$ tells you something, then $(\mathbf{f}_1, \mathbf{f}_2, ..., \mathbf{f}_n)$ should also tell you something, perhaps more than any one $\mathbf{f}_i$ does.

Interpretation. Classically, a system of n particles, each with 3 degrees of freedom, was described by a function $\emptyset$ of $6n$ variables which, for each specification of the $6n$ variables, gave the behavior or state of the system. If we choose a fixed 'time' set T and a fixed 'coordinate' set X, then we have a set $\{\mathbf{f}_1, ..., \mathbf{f}_{6n}\}$ of $6n$ functions from T into X and also the function $(\mathbf{f}_1, \mathbf{f}_2, ..., \mathbf{f}_n)$ from T into X^{6n} defined by

$$t \xrightarrow{\ (\mathbf{f}_1, ..., \mathbf{f}_{6n})\ } (\mathbf{f}_1(t), ..., \mathbf{f}_{6n}(t)) \in X^{6n}$$

Clearly, for each function $\mathbf{F}$ with domain X^{6n} and codomain Y, we have $\mathbf{F} \cdot (\mathbf{f}_1, ..., \mathbf{f}_{6n})$ from T into Y.

If $\mathbf{j}$ denotes the identity function from T into T which is given by $t \xrightarrow{\ \mathbf{j}\ } t$, then we have the function g defined by the diagram below whenever $\mathbf{F}$ is a function from X^{6n+1} into Y.

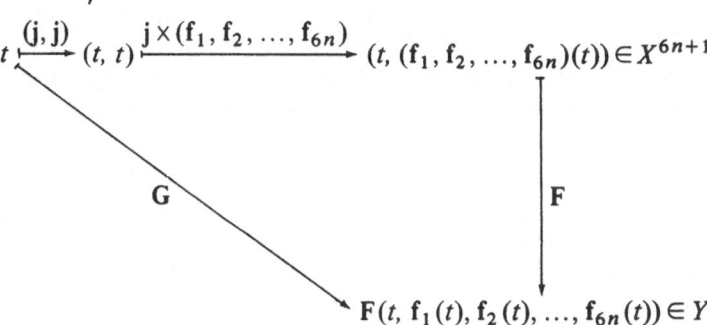

Notice that, for each $t \in T$, we have

$$\mathbf{G}(t) = \mathbf{F}(t, \mathbf{f}_1(t), ..., \mathbf{f}_{6n}(t)) \in Y$$

and $\mathbf{G}(t)$ in some sense gives the state of the system modeled by $\{\mathbf{f}_1, \mathbf{f}_2, ..., \mathbf{f}_{6n}\}$ and $\mathbf{F}$. Hence, by Axiom B, $\mathbf{G}$ is a state function for the object if $\mathbf{F}$ is a state function for θ.

Another interpretation would be the usual state functions of Quantum Mechanics. In both these examples, chosen from Physics, there are a number of restrictions on the state functions — some are purely technical and some are required by the Physics involved. One of the latter requirements is that the functions satisfy the dynamical stipulations chosen — in a number of situations, they are required to satisfy particular differential equations. Because we are aiming for generality, we are not now concerned with the specific form of the constraints on the state function.

Axiom D. If $\mathbf{f}: T \to X_1 \times X_2 \times \ldots \times X_n$ is a state function for θ, then $\Pi_i \cdot \mathbf{f} = \mathbf{f}_i$ is a state function for θ for $i = 1, 2, \ldots, n$.

Comment. If $\mathbf{f}$ tells you something, then each $\mathbf{f}_i$ should tell you something – perhaps less than $\mathbf{f}$ did but still something.

As proposed above, state functions are too general to allow us to do much with them. We need some kinds of restrictions on state functions and the most obvious way to proceed would be to require that the state functions satisfy some sort of change condition. After all, if all schemes really are schemes of change, then this feature should be reflected in some stipulation about the state functions. In an earlier example, our change condition required the function representing location to be twice-differentiable (and, implicitly, to have a continuous second derivative function). Precisely what form the change conditions might take will depend on the change being modeled, which also influences the choice of state functions. Two of the well-known change conditions are those involving differential or difference equations; less familiar are the change conditions involved with the Markov chain transition probabilities. Change conditions will sometimes depend on the state functions, their domains and codomains. Special domains and codomains allow special change conditions; differential change conditions don't exist on discrete domains. In any event, by a suitable choice we obtain the state function(s) and the change condition(s) for a scheme and we can turn to the question of law(s).

We return to our example from Mechanics. In that case we identified the law as the statement: $m\mathbf{D}^{(2)}\mathbf{x}(t) = \mathbf{F}(t)$ for all $t \in \mathbb{R}^+$ with $\mathbf{x}$ the state function and $\mathbf{F}$ the force function for the object (= the particle) whose motion is 'caused' by $\mathbf{F}$. In this case $\{t: m\mathbf{D}^{(2)}\mathbf{x}(t) = \mathbf{F}(t)\} = \mathbb{R}^+$ and so we have the situation described pictorially by

$$(\mathbf{x}, \mathbf{F}) \mapsto \{t: m\mathbf{D}^{(2)}\mathbf{x}(t) = \mathbf{F}(t)\}$$

which makes sense for each twice-differentiable function $\mathbf{x}$ and each function $\mathbf{F}$; we are not now concerned with whether we can obtain $\mathbf{x}$, given $\mathbf{F}$, only in the fact that the situation makes sense. Two facts should be observed. The first is that, depending on the pair of functions $(\mathbf{x}, \mathbf{F})$, the set on the right may be all of $\mathbb{R}^+$, the domain of both functions, or only a 'small' portion of the domain — 'small' in some suitable sense. The second is that we have a function H whose domain is $\mathscr{D}^{(2)} \times \mathscr{F}$ where $\mathscr{F}$ is the set of all functions from $\mathbb{R}^+$ into $\mathbb{R}$ and $\mathscr{D}^{(2)}$ is the subset of $\mathscr{F}$ consisting of all twice-differentiable functions and whose range consists of subsets of the common domain of the functions. Explicitly

$$(\mathbf{x}, \mathbf{F}) \xmapsto{\;\;\mathbf{H}\;\;} \mathbf{H}(\mathbf{x}, \mathbf{F}) \subset \mathbb{R}^+$$

and one seems to be saying, in this example, that for suitable functions $\mathbf{F}$, there exist twice-differentiable functions $\mathbf{x}$ such that $\mathbf{H}(\mathbf{x}, \mathbf{F}) = \mathbb{R}^+$ for our special function $\mathbf{H}$. It is not clear, even on classical grounds, which functions $\mathbf{F}$ can be 'force' functions. Of course, it is clear that here the state function $\mathbf{x}$ must be twice-differentiable but otherwise we know little about such state functions $\mathbf{x}$ and $\mathbf{Dx}$.

There is another function lurking in the background, namely the function described by the diagram

$$(\mathbf{x}, \mathbf{F}) \mapsto m\mathbf{D}^{(2)}\mathbf{x} - \mathbf{F}$$

which sends the pair of functions $(\mathbf{x}, \mathbf{F})$ from $\mathscr{D}^{(2)} \times \mathscr{F}$ to the function $m\mathbf{D}^{(2)}\mathbf{x} - \mathbf{F}$ in $\mathscr{F}$. If we call this function $\mathbf{L}$, then $\mathbf{L}(\mathbf{x}, \mathbf{F})$ is the function $m\mathbf{D}^{(2)}\mathbf{x} - \mathbf{F}$ and in our context

$$
\begin{aligned}
\mathbf{H}(\mathbf{x}, \mathbf{F}) &= \{t\colon m\mathbf{D}^{(2)}\mathbf{x}(t) - \mathbf{F}(t) = 0\} \\
&= \{t\colon \mathbf{L}(\mathbf{x}, \mathbf{F})(t) = 0\} = \mathbb{R}^+
\end{aligned}
$$

for suitable functions $\mathbf{F}$. For unsuitable functions $\mathbf{F}$, it may happen that $\mathbf{H}(\mathbf{x}, \mathbf{F}) \neq \mathbb{R}^+$, indeed, that $\mathbf{H}(\mathbf{x}, \mathbf{F})$ is a rather meager subset of $\mathbb{R}^+$. If $\mathbf{L}$ is to capture a notion of law, then for suitable pairs $(\mathbf{x}, \mathbf{F})$ of functions, the set $\mathbf{H}(\mathbf{x}, \mathbf{F})$, which depends on $\mathbf{L}$, should 'fill up' $\mathbb{R}^+$ in some appropriate sense. In a symbolic phrase, $\mathbf{H}(\mathbf{x}, \mathbf{F})$ should be 'of Category II'; in topological terms, $\mathbf{H}(\mathbf{x}, \mathbf{F})$ should not be the countable union of 'nowhere dense' sets. Exactly what 'Cat II' should mean would depend on a more explicit formulation of these considerations – the notion need not be topological.

Notice that in our case $\mathbf{L}$ acts on a pair of functions: in a relativistic setting, $\mathbf{L}$ would act on triples of functions $(\mathbf{m}, \mathbf{x}, \mathbf{F})$ and $\mathbf{L}(\mathbf{m}, \mathbf{x}, \mathbf{F})$ would then be the function $m\mathbf{D}^{(2)}\mathbf{x} - \mathbf{F}$ whose value at $t \in \mathbb{R}^+$ would be the real number $m(t)\mathbf{D}^{(2)}\mathbf{x}(t) - \mathbf{F}(t)$. In one instance, $\mathbf{L}$ acts on families of two functions; in the next instance, it acts on families of three functions.

We now proceed with the general framework. Ideally, we should have ten more spectacular motivating examples before we begin but space does not allow. We let $\tilde{\mathbf{F}}$ be a function from a nonempty set T such that, for each $t \in T$, $\tilde{\mathbf{F}}(t) = \tilde{\mathbf{F}}_t$ is a family $\{f_i\colon i \in I\}$ of functions such that $f_i\colon X_i \to Y_i$. If each X_i is the base set of a structure, let X be the base set of a canonical over-structure. (If each X_i is just a set, let X be their union. If vector spaces, then take, say, the direct product of the spaces, etc.) Similarly for Y. Then each function f_i is a nontrivial partial function from X to Y. Let $\mathbb{P}(X, Y)$ be the set of all partial functions from X into Y, and let $\mathscr{F}(\mathbb{P}(X, Y))$ be the set of all families of partial functions from X to Y. We assume that Y has a distinguished element, say, a 'zero'.

If $\mathscr{R}_{\widetilde{F}}$ denotes the range of $\widetilde{F}$, then suppose that L is a function whose domain includes $\mathscr{R}_{\widetilde{F}}$. Explicitly, we want

$$T \xrightarrow{\widetilde{F}} \mathscr{F}(\mathbb{P}(X, Y))$$

$$\mathscr{R}_{\widetilde{F}} \xrightarrow{L} \mathbb{P}(X, Y)$$

We shall then say that L is *extensive for* $\widetilde{F}$ *at* $t \in T$ if the set $\{x: x \in X, (L \circ \widetilde{F}_t)(x) = 0\}$ is of 'Cat II' in X. The function L is *a law for* $\widetilde{F}_t$ *on* X *and* Y if and only if L is extensive for $\widetilde{F}$ at $t \in T$. (If we can regard X and Y as determined, in some standard fashion, by the function $\widetilde{F}$, then we have defined the notion of a law for the image $\widetilde{F}_t$ of $t \in T$ under $\widetilde{F}$.)

We make some comments about this notion. First observe that $\widetilde{F}$ may be a constant function, that is, $\widetilde{F}_t$ for each $t \in T$ may be a fixed function $f: X \to Y$. $L \circ \widetilde{F}_t$ is then a fixed function $L(f)$ and we want the set $\{x: x \in X, L(f)(x) = 0\}$ to be 'large' as compared to X if L is to be a law for f. Again, $\widetilde{F}$ may always give the same family of two functions – recall our example from classical mechanics! For many situations $\widetilde{F}_t$ will be a finite family for each $t \in T$ although nothing in our formulation requires such a restriction.

Notice that L may be a law for some $\widetilde{F}_t$ and not for others. This is intended to allow that, say, L is not a law for $\widetilde{F}_1$, which is a pair of functions, but is a law for $\widetilde{F}_2$, which is a triple of functions. One could say that L is *a law for* $\widetilde{F}$ iff L is a law for at least one $\widetilde{F}_t$. Clearly, we only want to deal with laws for $\widetilde{F}$.

The situation above can be described in another way. For an object θ, we can look at the function $\widetilde{F}$ which, for each $t \in T$, provides a family of functions some or all members of which may be state functions for θ. And for a given function $\widetilde{F}$ we can look for laws for $\widetilde{F}$. But we could turn things around. For given structures X and Y with $0 \in Y$, we can look at functions L (perhaps only partial functions) which send families in $\mathbb{P}(X, Y)$ to single functions in $\mathbb{P}(X, Y)$. Such functions L will be possible laws in (X, Y). A particular L, of course, would be a law in (X, Y) iff there exists a function $\widetilde{F}$

such that, for some $t \in T$, $\{x : x \in X, (\mathbf{L} \cdot \tilde{\mathbf{F}}_t)(x) = 0\}$ is Cat. II in some appropriate sense. In this view, the L's are given and we search for the $\tilde{\mathbf{F}}$'s; in the other view, $\tilde{\mathbf{F}}$ is given and the L's are sought. In practice, because of the verification between realizations and schemes, both views are utilized.

The approaches sketched above also allow for the situation frequently found in Science; one law holds under centain circumstances or in one region of space and a different law holds under changed circumstances or in a different region of space – contrast classical and relativistic mechanics. Our approach allows for an arbitrary number of laws relative to a particular object.

We can now say that to each object θ we can correspond a dynamical system of the following form:

$$(\text{state function(s), change condition(s), law(s)})$$

with the provision that we may not have change conditions or laws. This latter provision would take care of the case of the stone lying buried in the earth. It has certain properties which could be described by a state function which, in essence, lists or measures these properties. Over the short run, probably no interesting change occurs, hence no change conditions. The laws here are constant. Hence, for the case of the stone, our dynamical system approach reduces to the usual notion of tabulated properties.

Lest the point of the exercise be lost, let me be explicit. I feel that the dynamical system approach can provide a pleasant conceptual framework for the beginner and a useful working tool for the practitioner. I feel that it has enough breadth and flexibility to accommodate much of what occurs in the application of Mathematics to Science. Hopefully we can learn something about objects by studying their dynamical systems which live in the World of Mathematics.

6. QUESTIONS

Some obvious questions relative to future work should perhaps be noted. We list a few of them below.

Science (or at least scientists) talk about the 'laws of nature' and there are varying opinions about what they are, what they do, and what role they play in Science (and more generally in Epistemology). Our proposed 'laws' are lodged in dynamical systems which are mathematical models. In particular, our 'laws' are mathematical.

Question A. Are all 'laws of nature' mathematical? or at least mathematically stateable?

Does 'law' carry with it the notion of universal or nearly universal applicability? How extensive should "laws" be? In our context, we have:

Question B. How should the notion of Category II be formulated to capture the notion of 'filling-up' X?

Some have remarked that really existing objects are stable — were they not, perturbations would have caused them to go out of existence. Considerations of stability lead to:

Question C. Is it possible to precisely formulate a notion of a stable object or the notion of a stable dynamical system relative to allowable perturbations?

The obvious change conditions are equational. But Mathematics is primarily relational. Hence:

Question D. Can change conditions be formulated in relational terms? Are there natural mathematical models involving relational change structures? Are there relational rather than functional laws?

Objects frequently combine (in some sense) with other objects. One wonders how to deal with the questions of combination. And one asks:

Question E. If objects combine to produce objects, can we usefully model this combination of objects by combinations of associated dynamical systems?

Returning to an earlier point, I would like to raise again the question of unification. For the record, I am in favor of unification. In particular, I am interested in seeing substantial movement made toward a unification of Systems Theory, Physics, and Foundations of Physics. An overall unification would be highly desirable but probably is not obtainable. However, the three fields above might readily serve as a test case for unification. They seem to be sufficiently close that an attempt to achieve a unification would at least be feasible. The literature at present is not particularly encouraging but there are some signs which provide hope. The basic hope resides in the interchange of ideas and there appears to be a growing concern for broadening the interchange. May it continue to grow!

REFERENCES

Hartnett, W. E.: 1963, *Principles of Modern Mathematics*, Book 1, Harper, New York; reprinted 1966 by Scott, Foresman, Chicago.

Hartnett, W. E.: 1971, *Principles of Modern Mathematics*, Book 2, Scott, Foresman, Chicago.
Hartnett, W. E.: 1973, 'The CF/PMM Approach to Learning Mathematics', *Educational Studies in Mathematics* 5, 1–22.
Nalimov, V. V.: 1974, 'Logical Foundations of Applied Mathematics', *Synthesis* 27, 211–250.
Wigner, E. P.: 1960, 'The Unreasonable Effectiveness of Mathematics in the Natural Sciences', *Communications in Pure and Applied Mathematics* XIII.
Windeknecht, T. G.: 1971, *General Dynamical Processes: A Mathematical Introduction*, Academic Press, New York.

INDEX

Abstraction 171
Action routines 31
Activity array(s) 125
Activity matrix 131
Adaptation (of a system) 69
Adjoint of a functor 15
Aggregate (= conglomerate) 89
Analysis 171
Appearances 122
Applied Mathematics 1, 180
Arbib, M. A. ix, xi, xii, 1, 2, 14, 15, 16, 25, 26, 27, 30, 31, 35, 39, 40, 42, 51, 54, 61
Artifact 172
Artificial intelligence (= AI) 27
Ashby, W. R. 121, 157
Association 67
Associative 9
Athans, M. 2, 26
Attributes 121
Automaton 45
Axiom(s) 190

Bainbridge, E. S. xi, xii, 14, 26, 45, 51, 54, 60, 61
Bartlett, F. C. 33, 34, 42
Basic behavior 133
Basic slide-box metaphor 30
Basic time 123
Basic variables 157
Behavior (of an automaton) 46
Behavior (of a network) 49
Bifurcation (objects) 173
Blackburn, T. 64
Blum, J. 28, 43
Bobrow, L. S. 2, 14, 26
Boole, G. 7
Boylls, C. C. 28, 42
Brain theory (= BT) 27
Buckley, W. xiii, 63
Bunge, M. ix, xi, xii, 71
Burnam, C. ix
Burt, P. 28, 29, 30, 43

Canada Council ix, 95, 159
Canonical (transformations) 88
Carnot-Clausius principle 63

Catastrophes 19
Category 9, 10, 94
Category theory 1, 10, 190
Category II 193, 196
Causation 67
Characterization 29
Circuit diagram 46
Codomain (of a morphism) 10
Cofree dynamics 15
Cofree $\mathscr{L}$-object over B 15
Colby, K. M. 21, 26, 31, 43
Collins, J. 98
Competition and cooperation routines 32
Components 89
Computer simulation 99
Conceivable state space 84
Conceptual framework 178
Conceptual framework (for mathematical modeling) 188
Context 184
Continuous time systems 6
Control 68
Controlled systems 129
Control system (= control channel) 68
Control vector 3
Coordinate description 45, 58
Coordinatization 57
Coordinatized automaton 57
Craik, K. J. W. 33, 43
Cybernetics (= feedback theory) 99

Data-generation model 103
Data matrix 157
Defined time 123
Dev, P. 28, 29, 43
Dichotomic global property 74
Didday, R. L. 28, 29, 30, 31, 39, 43
d-masks, rectangular 148
Discrete time systems 6
Distinguishable states 55
Distinguishable variables 55
Distributive law (in a category) 23
Domain (of a morphism) 10
Dual representation (of a representation) 53
Dynamical system (= DS) 187
Dynamics 11

DYNAMO III 115
Dyn (X) 14
Dyn (X)-morphism (= dynamorphism) 14

Econometrics 101
Education 177
Ehrig, J. 14, 26
Entropy function 142
Environment (of a system) 138
Equivalent (events) 94
Ernst, G. W. 39, 43
Error 174
Estimation model 103
Euler-Lagrange equation 80
Event space 93
Evolution of state vectors over time 3
Explanation 184
Extended mask 140
Extension (of a segment) 140
Extensive for 194

Falb, P. L. 2, 26
Feedback theory (= cybernetics) 99
Feinstein, A. 142, 157
Feller, W. 155, 157
15¢ machine 7
Force(s) 169
Forgetful functor 15
Forrester, J. W. ix, xi, xii, 97, 100, 104,
    118, 119
Forrester, N. B. 97, 119
Foundations of Mathematics 1
Frame 30
Free dynamics 15
Free ℒ-object over B 15
FTC 51
Full coordinate description (of an autom-
    aton) 52
Full state description (of a network) 52
Functor 9, 10
Furth, H. G. 31, 43
Fuzzy activity array 135
Fuzzy category 22
Fuzzy machine 1, 17, 21, 24
Fuzzy resolution level 122
Fuzzy-set automaton 1, 21
Fuzzy-set machine 21
Fuzzy sets 20
Fuzzy variable 122

Generalization 179
Generalized momenta 2
Generalized positions 2

General property 75
General system 123
Generating behavior 134
Generic (objects) 173
Geschwind, N. 30, 43
Glansdorff-Prigogine 63
Global occlusion system 38
G-matrix 135
Goffman, E. 31, 43
Goguen, J. A. 14, 16, 21, 26
Gregory, R. L. 33, 43

Hamiltonian 2
Hanson, A. L. 27, 28, 29, 31, 43
Hartmanis, J. 50, 61
Hartnett, W. E. ix, xi, xiii, 95, 177, 178,
    179, 196, 197
Hebb, D. O. 159, 168
Hill, F. J. 27, 43
Hintikka, J. ix
Homeostasis 66
Homomorphism (of automata) 52
Homomorphism (of networks) 52
Hudson Symposium, Eighth ix, xii, 27

Individual property 75
Information flow function 49
Information flow diagram 48
Initial state 45
Input-matching routines 31
Input(s) 11, 45
Input variables 129
Instantaneous state 169
Intrinsic (description) 173
Invariant (state functions) 87

Jodrell Bank 184

Kalman, R. E. 2, 26, 51, 61
Kaufmann, A. 122, 157
Kilmer, W. L. 28, 29, 39, 43
Klir, G. xi, 121, 122, 123, 126, 131, 157
Kondratyev wave 105, 114
Krippendorff, K. 153, 157

Law for 194
Lawful event 90
Law statement 78
Lawful state space 85
Lawful state space of θ for f 190
Lawful transformations 87
Law function 79
Lawvere, F. W. 1, 26

Lagrangian dynamics 80
Learning 30
Left adjoint 15
Left-most sampling variable 134
Lehrman, D. S. 64
Linear machine 1, 6, 8
Local feature 28
Logic 1
Lorentz-covariant 87
Luria, A. R. 30, 43

MacCorquodale, K. 159, 160, 168
McCulloch-Pitts 7
McCulloch, W. S. 28, 29, 33, 43
McFarland, D. J. 64
MacKay, D. M. 33, 43
MacLane, S. 22, 26
Machines in a category 1, 3, 11, 12, 24
Manes, E. G. ix, xi, xii, 1, 2, 14, 15, 16, 22, 25, 26, 51, 61
Markov chain 19
Mask 131
Mass, N. J. 97, 119
Mathematical models 180
Mechanics, generalized 1, 17
Meehl, P. E. 159, 160, 168
MG (= morphogenesis = self-organization) 66
Milner, P. xi, 159
Minimal (automaton) 55
Minimal (network) 55
Minsky, M. L. 30, 33, 43
Mistake 174
MIT 97
Modeling 182
Model(s) 101
Monkey-and-banana problem 39
Montalvo, F. S. 29, 43
Moore graph 91
Morphisms 9, 10
Morphogenesis (via amendments = MG_2) 66
Morphogenesis (= self-organization = MG) 66
Morphogenesis (via destructive conflict – MG_1) 66
Morphostasis (= self-stabilization = MS) 66
MS (= morphostasis = self-stabilization) 66

Nalimov, V. V. 182, 197
Nauta, W. J. H. 30, 43
Network 49
Neutral systems 129

Newell, A. W. 39, 43
Nondeterministic sequential machine 1, 16
Normalized (fuzzy activities array) 136
Normative theory of discovery 156

Object (concepts) 160
Object (of investigation) 121
Objects, familiar 30
Objects (in a category) 9, 10
Object (= studiable thing) 183
Object system 123
Observability 5
Observable 4, 54
Observable coordinate description (of an automaton) 55
Oltmans, W. L. 97, 119
Optimization 4
Orchard, R. A. 126, 131, 158
Organizing principles 179
Orlovsky, G. N. 28, 43
Output function 45
Outputs 45
Output variable(s) 49, 129

Padulo, L. 2, 14, 26
Pandemonium 28
Pattee, H. 175
Pergler, M. 65, 67
Peterson, G. R. 27, 43
Pfaffelhuber, E. 155, 158
Piaget, J. 33, 43
Probability 1, 17
Problem 1 (of Applied Mathematics) 181
Problem 2 (of Applied Mathematics) 181
Problems of control theory, three 4
Process 11, 89
Purpose (of investigation) 121

Qualitative global property 74
Quality Q_M of mask M 143
Quantitative global property 74
Quantitative global stochastic property 74
Quantitative local property 75

Reachability 5
Reachable 4, 54
Reachable state description (of a network) 55
Realization (of an automaton by a network) 52
Realization (of a network by an automaton) 53
Realization (of a scheme) 187

Realization problem 4, 5
Rectangular d-masks 148
Reduction of uncertainty 143
Reference (of a mask) 131
Regulation (via feedback) 68
Regulation (via measurement of the disturbance) 69
Relocation 29
Representation (of an automaton by a network) 52
Representation (of a network by an automaton) 53
Resolution level 31, 122
Response (concepts) 160
RETIC 32
Retrieves 50
Right adjoint 15
Right-most sampling variable 134
Riseman, E. M. 27, 28, 29, 31, 43
Rockefeller Brothers Fund 98
Rosenfeld, A. 30, 43
Rosen, R. xi, 169, 170, 173, 175

Sampling variables 131
Sangalli, A. A. L. 95
Schank, R. C. 21, 26, 31, 43
Schema 31
Schemes of things 183
Schutzenberger, M. P. 21, 26
Segmentation 29
Self-organization (= morphogenesis = MG) 66
Self-stabilization (= morphostasis = MS) 66
Semi-global features 28
Senge, P. M. 104, 119
Sequential machine 1, 7, 8
Selfridge, O. L. 28, 43
Set 8, 9
Set theory 1
Shannon, C. E. 142, 158
Simon, H. A. 41, 66, 156, 158
Source system 123
Space of events 91
Stable (set of variables) 59
Stateable (object) 190
State description 45, 57
State functions 75, 77, 190
State of θ at x (for f) 190
States 45, 122
State space of θ for f 190
State transition diagram 58
State transition function 45

State-transition relation 125
State-transition structure (= ST-structure) 133
State variables 75, 170
State vector 3, 77
Stearns, R. E. 50, 61
Stimulus (= excitation) 138
Stimulus-response (= S-R) 159
Stimulus-response-outcome 160
Stimulus-stimulus (= $S-S$) 159
ST-matrix 134
$S-R$ (= stimulus-response) 159
$S-S$ (= stimulus-stimulus) 159
Stochastic automaton 1, 20
Stravinsky, I. 121
Structurally stable system 18, 173
SUPERFROG 39
Supporting variables 157
System 69, 71, 89
System dynamics 98
System (level 0) 123
System(s) theory xii, 1
Szentágothai, J. 31, 43

Thom R. 19, 26, 173, 175
Three-dimensional activity array 135
Tolman, E. C. 159, 160
Traditional management 99
Transition congruence 59
Tree automaton 1, 13
Tuneability 30

Unification 178
United States 97

Values function 49
Variables 49, 122
Vect 9
Vision routines 31

Weaver, W. 142, 158
Wigner, E. P. 186, 197
Windeknecht, T. G. 190, 197
Winograd, S. 27, 43
Wymore, A. W. 51, 54, 59, 61

X-Machine 12

Young, J. Z. 33, 43

Zadeh, L. 20, 26, 122, 158
Zeiger, H. P. 51, 54, 61
Zeigler, B. P. 126, 127, 158